WITHDRAWN

Integrable Hamiltonian Systems on Complex Lie Groups

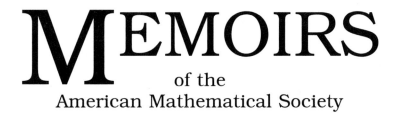

MEMOIRS
of the
American Mathematical Society

Number 838

Integrable Hamiltonian Systems on Complex Lie Groups

V. Jurdjevic

November 2005 • Volume 178 • Number 838 (second of 5 numbers) • ISSN 0065-9266

American Mathematical Society
Providence, Rhode Island

2000 *Mathematics Subject Classification.*
Primary 51N30, 49A25, 53C30, 53D05, 70S10.

Library of Congress Cataloging-in-Publication Data

Jurdjevic, Velimir.
  Integrable Hamiltonian Systems on Complex Lie Groups / V. Jurdjevic.
    p. cm. — (Memoirs of the American Mathematical Society, ISSN 0065-9266 ; no. 838)
  "Volume 178, number 838 (second of 5 numbers)."
  Includes bibliographical references.
  ISBN 0-8218-3764-8 (alk. paper)
  1. Hamiltonian systems. 2. Lie groups. 3. Manifolds (Mathematics). I. Title. II. Series.

QA3.A57   no. 838
[QA614.83]
510 s—dc22
[512′.482]   2005050811

Memoirs of the American Mathematical Society

This journal is devoted entirely to research in pure and applied mathematics.

Subscription information. The 2005 subscription begins with volume 173 and consists of six mailings, each containing one or more numbers. Subscription prices for 2005 are $606 list, $485 institutional member. A late charge of 10% of the subscription price will be imposed on orders received from nonmembers after January 1 of the subscription year. Subscribers outside the United States and India must pay a postage surcharge of $31; subscribers in India must pay a postage surcharge of $43. Expedited delivery to destinations in North America $35; elsewhere $130. Each number may be ordered separately; *please specify number* when ordering an individual number. For prices and titles of recently released numbers, see the New Publications sections of the *Notices of the American Mathematical Society.*
  Back number information. For back issues see the *AMS Catalog of Publications.*
  Subscriptions and orders should be addressed to the American Mathematical Society, P. O. Box 845904, Boston, MA 02284-5904, USA. *All orders must be accompanied by payment.* Other correspondence should be addressed to 201 Charles Street, Providence, RI 02904-2294, USA.
  Copying and reprinting. Individual readers of this publication, and nonprofit libraries acting for them, are permitted to make fair use of the material, such as to copy a chapter for use in teaching or research. Permission is granted to quote brief passages from this publication in reviews, provided the customary acknowledgment of the source is given.
  Republication, systematic copying, or multiple reproduction of any material in this publication is permitted only under license from the American Mathematical Society. Requests for such permission should be addressed to the Acquisitions Department, American Mathematical Society, 201 Charles Street, Providence, Rhode Island 02904-2294, USA. Requests can also be made by e-mail to `reprint-permission@ams.org`.

Memoirs of the American Mathematical Society is published bimonthly (each volume consisting usually of more than one number) by the American Mathematical Society at 201 Charles Street, Providence, RI 02904-2294, USA. Periodicals postage paid at Providence, RI. Postmaster: Send address changes to Memoirs, American Mathematical Society, 201 Charles Street, Providence, RI 02904-2294, USA.

Contents

Abstract

This paper is a study of the elastic problems on simply connected manifolds M_n whose orthonormal frame bundle is a Lie group G. Such manifolds, called the space forms in the literature on differential geometry, are classified and consist of the Euclidean spaces $\mathbb{E}^n$, the hyperboloids $\mathbb{H}^n$, and the spheres S^n, with the corresponding orthonormal frame bundles equal to the Euclidean group of motions $\mathbb{E}^n \rtimes SO_n(\mathbb{R})$, the rotation group $SO_{n+1}(\mathbb{R})$, and the Lorentz group $SO(1,n)$.

The manifolds M_n are treated as the symmetric spaces G/K with K isomorphic with $SO_n(R)$. Then the Lie algebra $\mathfrak{g}$ of G admits a Cartan decomposition $\mathfrak{g} = \mathfrak{p} + \mathfrak{k}$ with $\mathfrak{k}$ equal to the Lie algebra of K and $\mathfrak{p}$ equal to the orthogonal comlement $\mathfrak{k}$ relative to the trace form. The elastic problems on G/K concern the solutions $g(t)$ of a left invariant differential systems on G

$$\frac{dg}{dt}(t) = g(t)(A_0 + U(t)))$$

that minimize the expression $\frac{1}{2}\int_0^T (U(t), U(t))\, dt$ subject to the given boundary conditions $g(0) = g_0$, $g(T) = g_1$, over all locally bounded and measurable $\mathfrak{k}$ valued curves $U(t)$ relative to a positive definite quadratic form $(\,,\,)$ where A_0 is a fixed matrix in $\mathfrak{p}$.

These variational problems fall in two classes, the Euler-Griffiths problems and the problems of Kirchhoff. The Euler-Griffiths elastic problems consist of minimizing the integral

$$\frac{1}{2}\int_0^T \kappa^2(s)\, ds$$

with $\kappa(t)$ equal to the geodesic curvature of a curve $x(t)$ in the base manifold M_n with T equal to the Riemannian length of x. The curves $x(t)$ in this variational problem are subject to certain initial and terminal boundary conditions. The elastic problems of Kirchhoff is more general than the problems of Euler-Griffiths in the sense that the quadratic form $(\,,\,)$ that defines the functional to be minimized may be independent of the geometric invariants of the projected curves in the base manifold. It is only on two dimensional manifolds that these two problems coincide in which case the solutions curves can be viewed as the non-Euclidean versions of L. Euler elasticae introduced in 1744.

Each elastic problem defines the appropriate left-invariant Hamiltonian $\mathcal{H}$ on the dual $\mathfrak{g}^*$ of the Lie algebra of G through the Maximum Principle of optimal

Received by editor February 20, 2003; and in revised form May 18, 2004.

1991 *Mathematics Subject Classification.* 51N30, 49A25, 53C30, 53D05, 70S10

Key words and phrases. Lie groups, symmetric spaces, orthonormal frame bundles, elastic curves, optimal controls, Hamiltonian systems, integrability, co-adjointorbits control. The integral curves of the corresponding Hamiltonian vector field $\vec{\mathcal{H}}$ are called the extremal curves.

The paper is essentially concerned with the extremal curves of the Hamiltonian systems associated with the elastic problems. This class of Hamiltonian systems reveals a remarkable fact that the Hamiltonian systems traditionally associated with the movements of the top are invariant subsystems of the Hamiltonian systems associated with the elastic problems.

The paper is divided into two parts. The first part of the paper synthesizes ideas from optimal control theory, adapted to variational problems on the principal bundles of Riemannian spaces, and the symplectic geometry of the Lie algebra $\mathfrak{g}$ of G, or more precisely, the symplectic structure of the cotangent bundle T^*G of G.

The second part of the paper is devoted to the solutions of the complexified Hamiltonian equations induced by the elastic problems. The paper contains a detailed discussion of the algebraic preliminaries leading up to $so_n(\mathbb{C})$, a natural complex setting for the study of the left invariant Hamiltonians on real Lie groups G for which $\mathfrak{g}$ is a real form for $so_n(\mathbb{C})$. It is shown that the Euler-Griffiths problem is completely integrable in any dimension with the solutions the holomorphic extensions of the ones obtained by earlier P. Griffiths. The solutions of the elastic problems of Kirchhoff are presented in complete generality on $SO_3(\mathbb{C})$ and there is a classification of the integrable cases on $so_4(\mathbb{C})$ based on the criteria of Kowalewski-Lyapunov in their study of the mechanical tops. Remarkably, the analysis yields essentially only two integrables cases analogous to the top of Lagrange and the top of Kowalewski. The paper ends with the solutions of the integrable complex Hamiltonian systems on the $SL_2(\mathbb{C}) \times SL_2(\mathbb{C})$, the universal cover of $SO_4(\mathbb{C})$.

CHAPTER 1

Introduction

The mathematical formalism G. Kirchhoff used to model the equilibrium configurations of a thin elastic bar in $\mathbb{R}^3$ subject to twisting and bending torques at its ends extends naturally to arbitrary Riemannian manifolds M_n and leads to a class of variational problems, called elastic, on the orthonormal frame bundle of M_n. This paper is a study of the elastic problems on manifolds M_n whose orthonormal frame bundle is a Lie group. Such manifolds, called the space forms in the literature on differential geometry, are classified and consist of the Euclidean spaces $\mathbb{E}^n$, the hyperboloids $\mathbb{H}^n$, and the spheres S^n, with the corresponding orthonormal frame bundles equal to the Euclidean group of motions $\mathbb{E}^n \rtimes SO_n(\mathbb{R})$, the rotation group $SO_{n+1}(\mathbb{R})$, and the Lorentz group $SO(1, n)$.

If G denotes any of the above Lie groups, then the elastic problems concern the solutions $g(t)$ of a left-invariant differential system

$$(i) \qquad \frac{dg}{dt}(t) = g(t)(A_0 + \sum_{i=1}^{m} u_i(t)A_i)$$

that minimize the expression $\frac{1}{2}\int_0^T (u(t), Qu(t))\, dt$ subject to the given boundary conditions $g(0) = g_0$, $g(T) = g_1$, where $A_0, \ldots A_m$ are given matrices in the Lie algebra $\mathfrak{g}$ of G and where Q is a positive definite $m \times m$ matrix.

Each elastic problem can be naturally viewed as an optimal control problem with $u(t) = (u_1(t) \ldots u_m(t))$ playing the role of the control functions. The Maximum Principle of optimal control then identifies the appropriate left-invariant Hamiltonian $\mathcal{H}$ on the dual $\mathfrak{g}^*$ of the Lie algebra of G such that each optimal curve $g(t)$ is the projection of an integral curve $\xi(t)$ of the associated Hamiltonian vector field $\vec{\mathcal{H}}$.

The paper is essentially concerned with the solution curves of the Hamiltonian differential systems that correspond to the elastic problems. These systems fall in two classes, the Euler-Griffiths problem and the problem of Kirchhoff. The Euler-Griffiths elastic problem consists of minimizing the integral

$$\frac{1}{2}\int_0^T \kappa^2(s)\, ds$$

1

with $\kappa(t)$ equal to the geodesic curvature of a curve $x(t)$ in the base manifold M_n with T equal to the Riemannian length of x. The curves $x(t)$ in this variational problem are subject to certain initial and terminal boundary conditions (Definitions 1.1 and 1.2). The elastic problem of Kirchhoff (Definition 1.5) is more general than the problems of Euler-Griffiths in the sense that the quadratic form Q that defines the functional to be minimized may be independent of the geometric invariants of the projected curves in the base manifold. It is only on two dimensional forms that these two problems coincide. In this case the solutions curves can be viewed as the non-Euclidean versions of L. Euler elasticae introduced in 1744.

Apart from their primary interest in elasticity and geometry, the elastic problems show remarkable connections with the movements of the multidimensional tops. A detailed examination of the elastic problems in $\mathbb{R}^n$ reveals an even more remarkable fact: the Hamiltonian systems traditionally associated with the movements of the top are invariant subsystems of the Hamiltonian systems associated with the elastic problems. This observation has several significant consequences. First of all, it shifts the emphasis from mechanics to geometry and allows for non-Euclidean extensions of the theory of Hamiltonian systems associated with the spinning tops. Secondly, it properly accounts for the symmetries and justifies several ad-hoc practices in the integrability theory of Hamiltonian systems associated with the tops. Thirdly, and perhaps most importantly, it identifies a large class of Hamiltonian systems on Lie groups interesting in its own right and insightful for the geometry of the associated homogeneous spaces. Part of the motivation for this paper is to explain these claims in some detail.

The paper is divided into two parts. The first part of the paper synthesizes ideas from optimal control theory, adapted to variational problems on the principal bundles of Riemannian spaces, and the symplectic geometry of the Lie algebra $\mathfrak{g}$ of G, or more precisely, the symplectic structure of the cotangent bundle T^*G of G. The paper begins with the definitions of the elastic problems on the frame bundle of arbitrary Riemannian manifolds and then moves to the orthonormal frame bundle of the space forms (i.e., the frame bundle of simply connected manifolds of constant curvature)(Chapter I). Since the orthonormal frame bundles of space forms coincide with their isometry groups the focus shifts to Lie groups. If G denotes any one to those isometry groups then the corresponding space form can be realized as the quotient space G/K with K isomorphic with $SO_n(R)$. In particular each quotient G/K is identified with the orbit $\{ge_0 : g \in G\}$ where e_0 is the column vector in $\mathbb{R}^{n+1}$ whose first coordinate is equal to 1 and all other coordinates are equal to zero. In this representation the first column of g is identified with a point in the base space and the remaining columns are identified with an orthonormal frame at the this point. Elastic problems in this setting are certain variational problems defined over the absolutely continuous curves $g(t)$ in G, called Darboux curves that satisfy

$$\frac{dx}{dt} = \frac{d}{dt}(g(t)e_0) = g(t)e_1.$$

Alternatively, we may think of a Darboux curve as the most general lifting of a curve $x(t)$ in the base space to a curve $g(t)$ in the frame bundle adapted to the curve in such a way that the tangent vector $\frac{dx}{dt}$ coincides with the first vector of the frame defined by $g(t)$. Due to this constraint then each Darboux curve $g(t)$ is the solution of

$$\frac{dg}{dt}(t) = g(t)(E_1 + U(t))$$

with $U(t)$ an arbitrary curve in the Lie algebra $\mathfrak{k}$ of K and where E_1 is the matrix in $\mathfrak{g}$ such that $E_1 e_0 = e_1$. In the parlance of control theory, the left invariant vector field $g \to gE_1$ is called the drift.

It turns out that E_1 belongs to the orthogonal complement of $\mathfrak{k}$ relative to the trace form. It is shown in Chapter II that this orthogonal complement, denoted by $\mathfrak{p}$, satisfies the classic Lie algebraic relations:

$$[\mathfrak{p}, \mathfrak{p}] \subset \mathfrak{k}, [\mathfrak{p}, \mathfrak{k}] = \mathfrak{p}, [\mathfrak{k}, \mathfrak{k}] = \mathfrak{k}.$$

The decomposition of $\mathfrak{g}$ as the sum $\mathfrak{p} + \mathfrak{k}$, subject to the preceeding relations is known as the Cartan decomposition of $\mathfrak{g}$. It is known that each symmetric space G/K admits a Cartan decomposition. The generalized elastic problem concerns variational problems in arbitrary symmetric spaces G/K defined over differential systems (i) with A_0 a fixed element of $\mathfrak{p}$ and $A_1 \ldots A_m$ a basis of $\mathfrak{k}$.

Chapter III is devoted to the Maximum Principle and the Hamiltonians associated with control affine systems in an arbitrary manifold M_n. Such systems include the generalized elastic problems, all Riemannian problems and also all mechanical problems (through the Principle of Least Action).The Maximum Principle states that the optimal solutions are the projections of the extremal curves on the base manifold M_n, where the extremal curves are solution curves of certain Hamiltonian systems on the cotangent bundle of M_n. In contrast to the Riemannian and the mechanical problems, the elastic problems may be the projections of either the regular or the abnormal extremal curves.

Chapter IV deals with the left-invariant Hamiltonian systems on the cotangent bundle T^*G of a Lie group G realized as the product $G \times \mathfrak{g}^*$. Looking ahead to the second part of the paper this chapter provides a self contained treatment of the Poisson structure of the dual $\mathfrak{g}^*$ of a complex Lie algebra $\mathfrak{g}$ associated with an arbitrary complex Lie group G. The formalism may be seen as an extension of A. A. Kirrilov's coadjoint orbit theory to complex Lie groups. This material also includes the Hamiltonian equations on the Lie algebra $\mathfrak{g}$ both for the semi-simple case and the semi-direct products of Lie algebras.

The exposition then returns to the elastic problems and the related Hamiltonian systems on real Lie groups. In particular, Chapter IV contains a complete treatment of the abnormal extremals and sets the stage for the second part of the paper by showing that the Hamiltonian equations associated with mechanical tops in $\mathbb{R}^n$ form an invariant sub-system of the Hamiltonian equations associated with an elastic problem on the semi-direct product $\mathbb{R}^n \rtimes SO_n(\mathbb{R})$.

Chapter V provides a general background for the theory of integrable Hamiltonian systems on Lie groups. It is shown that all left-invariant Hamiltonian systems conform to certain invariants called either the conservation laws, or the Casimir functions. In general a function that remains constant along the flow of a Hamiltonian system is called an integral of motion, and a conservation law is an integral of motion valid for any left-invariant system. Chapter V contains a detailed study of the conservation laws. Remarkably, the conservation laws for the Euclidean group E_n may be obtained as the singular limit of the conservation laws for the semi-simple Lie groups SO_{n+1} and $SO(n,1)$. The significance of the conservation laws is then illustrated through the theorem of Poinsot concerning the movements of the top of Euler. This theorem, known exclusively as a theorem in mechanics, is equally applicable for the elastic problems in $\mathbb{E}^3$ where it appears as a consequence of the conservation laws on $se_3(\mathbb{R})$.

The second part of the paper is devoted to the solutions of the complexified Hamiltonian equations induced by the elastic problems. Chapter VI contains the algebraic preliminaries: an involutive automorphism on $SO_n(\mathbb{C})$ induces a Cartan decomposition $\mathfrak{g} = \mathfrak{k} + \mathfrak{p}$ on the Lie algebra $so_n(\mathbb{C})$ through which the complexified elastic Hamiltonians are defined. Then the vector space of $n \times n$ antisymmetric matrices with complex entries may be considered as a double Lie algebra, the semi direct product $\mathfrak{p} \rtimes \mathfrak{k}$ and the usual $so_{n+1}(\mathbb{C})$ Lie algebra. The real forms of these complex Lie algebras include the Lie algebras $so(p,q)$ with $p + q = n + 1$ and the various semi direct products $\mathbb{R}^n \rtimes so_n(\mathbb{R})$ and $\mathbb{R}^n \rtimes so(p,q), p + q = n$.

Hence, $so_n(\mathbb{C})$ is a natural complex setting for the study of the left invariant Hamiltonians on the corresponding real Lie groups G. With this observation in mind, the complex Lie algebras associated with $n = 4$ are of particular importance. There is a detailed discussion of $SO_4(\mathbb{C})$, its universal cover $SL_2(\mathbb{C}) \times SL_2(\mathbb{C})$ and the real forms of the corresponding Lie algebras.

Chapter VII presents the solutions to the complexified Euler-Griffiths problem. It is shown that this problem is completely integrable in any dimension, and the solutions, described by Proposition 7.1, are complex extensions of a theorem of P. Griffiths ([**Gr**]). More explicitly, it is shown that each elastic curve is confined to a three dimensional complex sub-manifold of $S^n = \{z \in \mathbb{C}^{n+1} : z_1^2 + \ldots z_{n+1} = 1\}$. The complex curvature $\kappa(t)$ of an elastic curve is a solution of

$$\frac{d\xi^2}{dt} = -\xi^3 + A\xi^2 + B\xi + C$$

for some constants A, B, C, and the complex torsion $\tau(t)$ associated with the elastic curve is determined through the relation $\kappa(t)^2 \tau(t) = $ constant.

The rest of the paper is devoted to the solutions of the complexified generalized Kirchoff's problems on six dimensional Lie groups that admit only meromorphic solutions of complex time. The investigation of the solutions begins with an analysis that is essentially due to A. M. Lyapunov ([**Ly**]) motivated by the results of S. Kowalewski in her original paper on the top published in 1889 ((([**Kw**]). Like the problems of the top which are described by six real parameters, the coordinates of the center of gravity and the principal moments of inertia, the generalized Kirchhoff's problems are described by b_1, b_2, b_3 and $\lambda_1, \lambda_2, \lambda_3$, with b_1, b_2, b_3 equal to the coordinates of the drift vector relative to a basis in the Cartan space $\mathfrak{p}$, and $\lambda_1, \lambda_2, \lambda_3$ equal to the eigenvalues of the quadratic form Q on $\mathfrak{k}$.

The classification of systems according to the Kowalewski-Lyapunov criteria form the content of Chapter VIII. Remarkably, the analysis yields the same results for the elastic problems as it did in the case of the top with only two cases that admit meromorphic solutions:

1. $b_1 = b_2 = 0$ and b_3 arbitrary, $\lambda_1 = \lambda_2$, and λ_3 arbitrary. Analogous to the top of Lagrange, $M_3 = $ constant and the system is completely integrable.

2. $b_3 = 0$ and b_1 and b_2 arbitrary, and $\lambda_1 = \lambda_2 = 2\lambda_3$. As in the case of the top of Kowalewski there is an extra integral of motion $I_4 = (z_1^2 - a(w_1 - \epsilon a)(z_2^2 - \bar{a}(w_2 - \epsilon \bar{a})$. This integral of motion is a holomorphic extension of the integral of motion reported in ([**Ju3**] and the variables which define the above integral correspond to the root space decomposition of $SO_4(\mathbb{C})$ induced by a certain Cartan algebra on $so_4(\mathbb{C})$.

Chapter IX provides the solutions for the case of Lagrange. The Hamiltonian equations are integrated on $SL_2(\mathbb{C}) \times SL_2(\mathbb{C})$ which is a double cover of $SO_4(\mathbb{C})$. Apart from getting the explicit solutions, the integration procedure reveals that the solutions on each of the groups in question (i.e., $\mathbb{C}^3 \rtimes SO_3(\mathbb{C})$, $SO_4(\mathbb{C})$ and $SL_2(\mathbb{C}) \times SL_2(\mathbb{C})$) are indeed meromorphic functions of complex time.

Chapter X, the last chapter of the paper, deals with the elastic problems of Kowalewski type. The first part of the chapter is directed to providing conceptual explanations for the existence of extra integrals of motion under the " mysterious" conditions $\lambda_1 = \lambda_2 = 2\lambda_3$ and $b_3 = 0$. The clue to the answer lies in the paper of I.A. Bobenko et al ([**Bb**]) in which Kowalewski's integral of motion obtained for the gyrostat in two constant fields appears as an isospectral invariant of a system in the Lie algebra Sp_4.

This paper provides an original interpretation of the results obtained by the authors cited above by demonstrating, without any reference to the R-matrices, that the integral of motion obtained by S. Kowalewski is a consequence of certain symmetries in $SO_5(\mathbb{C})$ which is shared with the semi-direct product $\mathbb{C}^3 \rtimes SO_3(\mathbb{C})$ and its real form $\mathbb{E}^3 \rtimes SO_3(\mathbb{R})$. The symmetry in $so_5(\mathbb{C})$ explains the mysterious conditions under which Kowalewski's integral of motion appears, and also provides natural explanations for the Lax pair representations and their integrals of motions obtained for the gyrostat in two constant fields ([**Bb**]). The paper ends with the resolution of the Hamiltonian equations in terms of the famous hyperelliptic curve of Kowalewski obtained through the change of coordinates based on the addition formulas of A. Weil.

This study completes the material presented in ([**Ju2**]), ([**Ju3**]) and ([**Gr**]. It contains several new results, the most fundamental of which is the holomorphic extension of the integral of motion discussed in ([**Ju3**]). The discovery of this holomorphic integral of motion shifts the focus to complex Lie groups as the natural setting for the classification of integrable cases based on the Lyapunov-Kowalewski criteria. The classification of integerable cases in the present paper improves the results reported in ([**Ju3**]. Furthermore, the integration procedure used in Chapter IX is new.

Even though there are overlaps with the theory traditionally associated with mechanical tops, this paper is not about the mechanical tops. Ultimately, the paper is about complex Lie groups and their homogeneous spaces. It is a study of a particular class of complex Hamiltonian systems on complex Lie groups whose real forms provide solutions to problems of Riemannian geometry and applied mathematics. Seen from this perspective, the paper advocates systematic use of complex symplectic geometry in the study of Lie groups and their homogeneous spaces.

I. Elastic Curves

The mathematical formalism G. Kirchhoff used to model the equilibrium configurations of a thin inextensible elastic rod in $\mathbb{R}^3$ subject to bending and twisting torques at its ends extends naturally to an arbitrary Riemannian manifold M_n, and leads to a class of variational problems on the oriented orthonormal frame bundle of M_n that is fundamental for this paper. To put this matter in proper perspective, it might be instructive to recall Kirchhoff's contributions to the theory of elastic rods.

According to A.E.Love , Kirchhoff considered an elastic rod as a framed curve in $\mathbb{R}^3$ ([**Lv**]). As such, the rod is represented by a parametrized curve $\gamma(t), 0 \le t \le T$ that corresponds to the central line of the rod, and an orthonormal frame $F(t) = (v_1(t), v_2(t), v_3(t))$ defined along $\gamma(t)$ that measures the amount of bending and twisting of the rod. To model inextensible rods, it is assumed that $\|\frac{d\gamma}{dt}(t)\| = 1$ which implies that the parameter t corresponds to the length of γ in the interval $[0, t]$ measured from the initial point $\gamma(0)$. The frames $F(s)$ are assumed adapted to the central line by a constraint of the form

$$\frac{d\gamma}{dt}(t) = v_1(t).$$

It is implicitly assumed that the frames $F(s)$ are measured relative to a fixed reference frame defined by a distinguished framed curve $\sigma_0(t)$ that corresponds to the rod in its unstressed state. Then the orthonormal frames can be identified with elements of $SO_3(\mathbb{R})$ in which case the derivative $\frac{dF}{dt}(t)$ is described by an antisymmetric matrix

$$A(t) = \begin{pmatrix} 0 & -u_3(t) & u_2(t) \\ u_3(t) & 0 & -u_1(t) \\ u_2(t) & u_1(t) & 0 \end{pmatrix}$$

that corresponds to the deformation of the rod. The functions $u_1(t), u_2(t), u_3(t)$ are called strains, which together with constants c_1, c_2, c_3 reflect the tensile and geometric properties of the rod defining the total elastic energy E of the deformed rod given by

$$E = \int_0^L (c_1 u_1^2(t) + c_2 u_2^2(t) + c_3 u_3^2(t)) \, dt.$$

In 1859 Kirchhoff postulated that the equilibrium configurations of the rod subjected to fixed boundary conditions at its ends correspond to stationary configurations for the elastic energy ([**Lv**],Chapter VII). In the absence of conjugate points, Kirchhoff's principle can be paraphrased as a Minimum Principle:

The equilibrium configurations of the rod correspond to the minima of the elastic energy relative to the configurations that conform to the given boundary conditions.

The passage from the planar case, initiated and solved by L. Euler in 1744, to the spatial case of Kirchhoff in 1858 marks a conceptual geometric difference. For framed curves $\sigma(s) = (\gamma(s), F(s))$ in the plane that satisfy

$$\left\| \frac{d\gamma}{dt}(t) \right\| = 1 \,, \frac{d\gamma}{dt}(t) = v_1(t)$$

the deformations of the frame are measured by the geodesic curvature κ of γ and the elastic energy is a constant multiple of $\frac{1}{2} \int_0^L \kappa^2(s)\, ds$. Therefore, the elastic problem of Euler is fundamentally a geometric problem in the sense that it can be formulated entirely in terms of the geometric invariants of the curve γ.

However, for spatial curves there may be no connection between the geometric invariants of the central line γ and the elastic problem formulated by Kirchhoff. To begin with the Serret-Frenet frames are inadequate for describing the equilibrium configurations of elastic rods, as first noticed by Saint-Venant ([**Lv**],Chapter XIV) (for instance, a twisted straight rod cannot be described by the Serret-Frenet frames). Even for curves that admit a lifting to the frame bundle via the Serret-Frenet frames there may not be any natural relation between the curvature and the torsion of the curve and the elastic energy functional.

Nevertheless, the geometric problem of minimizing the integral $\frac{1}{2} \int_0^L \kappa^2(s)\, ds$ over the curves $\gamma(s)$ in $\mathbb{R}^n$ is a natural mathematical problem independently of the connections to the elastic problem of Kirchhoff. It may be regarded as a prototype of a more general class of variational problems defined by the geometric invariants of the curve $\gamma(s)$. Of course these variational problems are only well defined relative to some prescribed class of admissible curves that conform to some fixed boundary conditions. The most natural boundary conditions for the problem of minimizing the integral $\frac{1}{2} \int_0^L \kappa^2(s)\, ds$ prescribe the position and the tangent vector at each extremity of the curve γ ([**Gr**]), but the problem could also be defined as a variational problem in the orthonormal frame bundle of $\mathbb{R}^3$ in which the admissible curves match the prescribed frames at their extremities. The nuance in the formulation of these geometric problems and their implications to the solutions will be addressed in more detail later on in the paper.

With these introductory remarks in mind let us consider a class of variational problems in arbitrary Riemannian manifolds M_n that are inspired by these elasto-geometric problems in $\mathbb{R}^2$ and $\mathbb{R}^3$. To do so it will be necessary first to introduce relevant notations related to the mathematical foundations.

As usual $T_x(M_n)$ and $T_x^*(M_n)$ denote the tangent and the cotangent space at $x \in M_n$, while $T(M_n)$ and $T^*(M_n)$ denote the tangent and the cotangent bundle of M_n . For each pair of tangent vectors u and v, the Riemannian scalar product will be denoted by $\langle u, v \rangle$ with $\|v\| = \sqrt{\langle v, v \rangle}$.

The covariant derivative of a curve of tangent vectors $v(t)$ defined along a curve $\gamma(t)$ in M_n will be denoted by $\frac{D_\gamma}{dt}(v)$. If $u(t)$ and $v(t)$ are vectors defined along a curve $\gamma(t)$ then the following holds ([**dC**])

(1) $$\frac{d}{dt}\langle u(t), v(t)\rangle = \langle \frac{D_\gamma}{dt}(u)(t), v(t)\rangle + \langle u(t), \frac{D_\gamma}{dt}(v)(t)\rangle$$

The positively oriented orthonormal frame bundle of M_n will be denoted by $\mathcal{F}_+(M_n)$, while the projection map from $\mathcal{F}_+(M_n)$ onto M_n will be denoted by π. The frame bundle $\mathcal{F}_+(M_n)$ is a principal $SO_n(R)$ bundle over M_n and therefore, parametrized curves $\sigma(t)$ in $\mathcal{F}_+(M_n)$ can be represented by pairs $(\gamma(t), F(t))$ with $\gamma(t)$ equal to the projection of σ in M_n and $F(t) = (v_1(t) \ldots v_n(t))$ the orthonormal frame along γ ([**St**]).

DEFINITION 1.1. *The set of framed curves $\sigma(t)$ defined over a fixed interval $[0, T]$ for which $\gamma(t) = \pi((\sigma(t))$ satisfies:*

$$\frac{d\gamma}{dt}(t) = v_1(t)$$

shall be called Darboux curves.

It follows that $\|\frac{d\gamma(t)}{dt}(t)\| = 1$, and therefore T is the Riemannian length of γ measured from the initial point $\gamma(0)$.

For each Darboux curve $\sigma \in \mathcal{F}_+(M)$ the deformations of the associated frame $F(t) = (v_1(t), \ldots, v_n(t))$ will be described by the matrix $U(t) = (U_{ij}(t))$ via the relations

(2) $$\frac{D_\gamma v_i(t)}{dt} = \sum_{j=1}^{n} U_{ij}(t)v_j(t)$$

for $i = 1, 2, \ldots, n$. It will be convenient to write $\frac{DF}{dt}(t) = F(t)U(t)$ for the equation (2)

It follows from (1) that $U(t)$ is an anti-symmetric matrix. . Moreover,

(3) $$\|\frac{D_\gamma}{dt}(\frac{d\gamma}{dt})\|^2 = \sum_{j=1}^{n} U_{1j}^2(t).$$

Since the geodesic curvature κ of γ is equal to $\|\frac{D_\gamma}{dt}(\frac{d\gamma}{dt})\|$ the variational problem of Euler can be formulated in terms of the constrained Darboux curves as follows.

DEFINITION 1.2 (EULER-GRIFFITHS PROBLEM 1). *Find a Darboux curve that minimizes the integral $\frac{1}{2}\int_0^T (\sum_{j=1}^n (U_{1j}(t))^2)\, dt$ over all Darboux curves $\sigma(t) = (\gamma(t), F(t))$, $t \in [0, T]$ that*

(3) *(i) are the solutions of*

$$\frac{d\gamma}{dt}(t) = v_1(t), \frac{DF}{dt}(t) = F(t)U(t)$$

with $U(t)$ antisymmetric matrices with bounded and measurable entries in the interval $[0,T]$ constrained by $U_{ij}(t) = 0$ for $i > 1, j \geq 1$,

and in addition

(4) *(ii) satisfy the boundary conditions $\gamma(0) = \gamma_0$, $F(0) = F_0$ and $\gamma(T) = \gamma_1$, $F(T) = F_1$.*

The projection $\gamma(t)$ of any Darboux curve that meets the boundary conditions *(ii)* in the Euler-Griffiths problem not only has a fixed position and a fixed tangent vector at the initial and the terminal point but it also conforms to the additional constraints imposed by the given frames F_0 and F_1.

To show continuity with the existing results in the literature ([**Gr**]) it will be necessary to consider the case where these higher order constraints on γ are not present. In such a case the boundary conditions *(ii)* in the above definition should be replaced by $\sigma(0) \in S_0$ and $\sigma(T) \in S_1$ where S_0 and S_1 are submanifolds of $\mathcal{F}_+(M)$ determined by the first order jets of γ

$$\gamma(0) = \gamma_0, \frac{d\gamma}{dt}(0) = \dot{\gamma}_0 \text{ and } \gamma(T) = \gamma_1, \frac{d\gamma}{dt}(T) = \dot{\gamma}_1,$$

i.e.,

$$S_0 = \{(x,F) \in \mathcal{F}_+(M) : x = \gamma_0, v_1 = \dot{\gamma}_0\}, S_1 = \{(x,F) \in \mathcal{F}_+(M) : x = \gamma_1, v_1 = \dot{\gamma}_1\}.$$

This new version will be stated as

DEFINITION 1.3 (EULER-GRIFFITHS PROBLEM 2). *Find a Darboux curve that minimizes the integral $\frac{1}{2}\int_0^T (\sum_{j=1}^n (U_{1j}(t))^2)\,dt$ over all Darboux curves $\sigma(t) = (\gamma(t), F(t))$, $t \in [0,T]$ that*

(5) *(i) are the solutions of*

$$\frac{d\gamma}{dt}(t) = v_1(t), \frac{DF}{dt}(t) = F(t)U(t)$$

with $U(t)$ anti-symmetric matrices with bounded and measurable entries in the interval $[0,T]$ constrained by $U_{ij}(t) = 0$ for $i > 1, j \geq 1$,

and also

(6) *(ii) satisfy the boundary conditions $\sigma(0) \in S_0$ and $\sigma(T) \in S_1$.*

The difference in the solutions of these two similar problems will be made clear by the subsequent exposition.

DEFINITION 1.4. *The projections* $\gamma(t) = \pi(\sigma(t))$ *of the solution curves* σ *of either of the above variational problems will be called the Euler-Griffiths elastic curves.*

The variational problems of Kirchhoff require another ingredient, a positive definite form in the space of deformations of Darboux curves that defines the "elastic energy" of a Darboux curve $\sigma(t) = \gamma(t), F(t))$, $t \in [0, T]$.

Let $\langle\langle U, V, \rangle\rangle_\sigma$ denote a smooth field of positive quadratic forms over Darboux curves in the space of $n \times n$ antisymmetric matrices U and V. Then

DEFINITION 1.5 (ELASTIC PROBLEM OF KIRCHHOFF). *Find a Darboux curve that minimizes the integral* $\dfrac{1}{2} \displaystyle\int_0^T \langle\langle U(t), U(t)\rangle\rangle_{\sigma(t)}\, dt$ *over all Darboux curves* $\sigma(t)$, $0 \le t \le T$ *that satisfy the conditions:*

(7) *(i)* $\sigma(0) = \sigma_0$ *and* $\sigma(T) = \sigma_1$ *where* σ_0 *and* σ_1 *denote the prescribed boundary conditions in* $\mathcal{F}_+(M)$, *and*

(8) *(ii)* $\dfrac{d\gamma}{dt}(t) = v_1(t)$, *and* $\dfrac{D_\gamma v_i}{dt}(t) = \displaystyle\sum_{j=1}^{n} U_{ij}(t) v_j(t)$, *where the functions* $U_{ij}(t)$ *are bounded and measurable in the integral* $[0, T]$.

DEFINITION 1.6. *The projections in* M *of the solutions to Kirchhoff's elastic problem will be called Kirchhoff's elastic curves.*

The above variational problems will be referred to as the elastic problems. Note that the elastic energy for these problems, the analogue of the Langrangian for problems of mechanics, is confined to the deformations of the frame and acts only indirectly on the curve through the constraints $\dfrac{d\gamma}{dt} = v_1(t)$. For that reason this class of variational problems falls outside of the conventional framework of mechanics. The relevance of this observation will become more transparent in the specific situations that follow.

1. Elastic problems on the space forms.

On manifolds whose orthonormal frame bundle is equal to the isometry group the elastic problems are naturally formulated as optimal control problems on Lie groups and their solutions are amenable by the theoretic methods developed in ([**Ju2**]). This paper is essentially motivated by the solutions of elastic problems in these spaces and their connections to to the equations of motion of mechanical systems. The manifolds in question can be defined as complete simply connected manifolds with constant sectional curvature, known as space forms in the literature on differential geometry.

There are only three simply connected space forms: The Euclidean space $\mathbb{E}^n$, the sphere S^n and the hyperboloid $\mathbb{H}^n$ ([**doC**]). The sections below describe the geometric foundation required for proper reformulations of the elastic problems in these spaces.

A. The Euclidean case.

Let $\mathbb{E}^n$ denote an n-dimensional Euclidean space with its inner product denoted by $\langle x, y \rangle$ for x and y in $\mathbb{E}^n$. Then O_n will denote the group that leaves the Euclidean inner product invariant, and SO_n will denote the connected component of O_n that contains the group identity. The semidirect product $\mathbb{E}^n \rtimes SO_n$ will be denoted by SE_n. Recall that SE_n consist of pairs $(x, R))$ with $x \in \mathbb{E}^n$ and $R \in SO_n$, and that the group operation is given by

$$(x, R)(y, S) = (x + Ry, R \circ S).$$

with $(0, I)$ the group identity and $(x, R)^{-1} = (-R^{-1}x, R^{-1})$ the group inverse.

The isomorphism between SE_n and the positively oriented orthonormal frame bundle $\mathcal{F}_+(\mathbb{E}^n)$ is obtained through the action of SE_n on $\mathbb{E}^n$ given by $(x, R)(y) = Ry + x$ for each element (x, R) of SE_n and each y in $\mathbb{E}^n$. This action extends to the frame bundle of $\mathbb{E}^n$ with $(v_1, \ldots, v_n) \to (Rv_1, \ldots, Rv_n)$.

Evidently the action of SE_n on the frame bundle of $\mathbb{E}^n$ is transitive with the isotropy group equal to the group identity. Hence the oriented orthonormal frame bundle can be identified with the orbit of SE_n through a fixed orthonormal frame $(e_1, \ldots, e_n)$ at the origin in $\mathbb{E}^n$. In this identification each framed curve $\sigma(t)$ is identified with a curve $(x(t), R(t))$ in SE_n. Then $x(t)$ is the projection of $\sigma(t)$ on $\mathbb{E}^n$ and $R(t)$ corresponds to the frame $F(t) = (v_1, \ldots v_n)$ via the relations $v_i(t) = R(t)e_i$, $i = 1, \ldots, n$. For Darboux curves the frame $F(t)$ is adapted to the curve $x(t)$ through the relation $\frac{dx}{dt}(t) = v_1(t)$.

Hence Darboux curves in Euclidean spaces $\mathbb{E}^n$ coincide with the solutions of

$$(4) \qquad \frac{dx}{dt}(t) = R(t)e_1 \quad \text{and} \quad \frac{dR(t)}{dt} = R(t)U(t)$$

B. The spherical and the hyperbolic cases .

The sphere S^n is a Riemannian manifold with its Riemannian metric inherited from the standard Euclidean metric in $\mathbb{R}^{n+1}$. To show that the oriented orthonormal frame bundle is isomorphic with $SO_{n+1}(\mathbb{R})$ consider the action of $SO_{n+1}(\mathbb{R})$ on $\mathbb{R}^{n+1}$ given by

$$(g, x) \rightarrow gx$$

where $x \in \mathbb{R}^{n+1}$ is taken as a column vector and where gx denotes the matrix multiplication by g. This action restriced to the sphere is transitive, and therefore the sphere can be realized as the homogeneous manifold SO_{n+1}/K with K equal to the isotropy group of a given point x_0 in S^n.

In particular when $x_0 = e_1$ then $K = \{1\} \times SO_n(\mathbb{R})$ and every point x of the sphere can be represented as $x = ge_1$ for some $g \in SO_{n+1}(\mathbb{R}^n)$. Since the columns of g are orthonormal relative to the standard inner product on $\mathbb{R}^{n+1}$, the remaining columns of g can be identified with an orthonormal frame $F = (v_1, \ldots v_n)$ at x via the relations

$$v_1 = ge_2 , \quad \ldots \quad , \quad v_n = ge_{n+1}.$$

This identification orients the orthonormal frames on S^n, and shows that the bundle of positively oriented orthonormal frames is equal to $SO_{n+1}(\mathbb{R})$. It follows that framed curves $\sigma(t)$ are described by curves $g(t)$ in $SO_{n+1}(\mathbb{R})$ with $x(t)e_1$ equal to the projection $\gamma(t)$ of $\sigma(t)$ on S^n.

The covariant derivative $\dfrac{D_x v}{dt}(t)$ of a curve $v(t)$ of tangent vectors along a curve $x(t)$, equal to the orthogonal projection of the ordinary derivative $\frac{dv}{dt}$ in $\mathbb{R}^{n+1}$ onto the tangent space $T_x S^n$, is given by the following expression:

$$\frac{D_x v}{dt} = \frac{dv}{dt} + \langle v(t), \frac{dx}{dt} \rangle x(t)$$

Then Darboux curves $g(t)$ satisfy

$$\frac{dx}{dt}(t) = v_1(t) = \frac{dg}{dt}(t)e_1 = g(t)e_2$$

$$\sum_{j=1}^{j=n} U_{ij}(t)v_j(t) = \frac{D_x v_i}{dt}(t) = \frac{dv_i}{dt}(t) + \langle v_i(t), \frac{dx}{dt}(t) \rangle x(t)$$

$$= \frac{dg}{dt}(t)e_{i+1} + \langle g(t)e_{i+1}, g(t)e_2 \rangle g(t)e_1$$

Therefore,

$$\frac{dg}{dt}(t)e_1 = g(t)e_2 \, , \, \frac{dg}{dt}(t)e_2 = -g(t)e_1 + \sum_{j=1}^{j=n} U_{1j}(t)g(t)e_{j+1},$$

and

$$\frac{dg}{dt}(t)e_i = \sum_{j=1}^{j=n} U_{ij}(t)g(t)e_{j+1}, \text{ for } i \geq 2.$$

The preceding system of equations can be written more compactly as:

$$(5) \qquad \frac{dg(t)}{dt} = g(t) \begin{pmatrix} 0 & -1 \cdots 0 \\ 1 & \\ \vdots & U(t) \\ 0 & \end{pmatrix}.$$

Thus Darboux curves in $\mathcal{F}_+(S^n)$ coincide with the solutions of (6) for arbitrary anti-symmetric matrices $U(t)$ with bounded and measurable entries in an interval $[0, T]$.

The description of Darboux curves in the frame bundle of the hyperboloid $\mathbb{H}^n = \{(x_1, \ldots x_{n+1}) : x_1^2 - \sum_{i=2}^{n+1} x_i^2 = 1, \ x_1 > 0\}$ is practically the same as that on the sphere. The Riemannian metric on $\mathbb{H}^n$ is inherited from the Lorentzian inner product in $\mathbb{R}^{n+1}$

$$\langle x, y \rangle = -x_1 y_1 + \sum_{i=2}^{n+1} x_i y_i.$$

The subgroup of $GL_{n+1}(\mathbb{R})$ that leaves the Lorentzian inner product invariant is denoted by $O(1, n)$. It is easily shown that the determinant $Det(M) = \pm 1$ for any element of $O(1, n)$. Then $SO(1, n)$ denotes the subgroup of $O(1, n)$ of matrices M whose determinant is equal to 1. The connected component of $SO(1, n)$ that contains the identity will be denoted by $SO_0(1, n)$.

Similar to the spherical case, the oriented orthonormal frame bundle of $\mathbb{H}^n$ is identified with $SO_0(1, n)$ via the relations

$$x = ge_1 \, , v_i = ge_{i+1} \, , i = 2, \ldots, n.$$

Then Darboux curves are identified with curves $g(t)$ in $SO_0(1, n)$ that satisfy $\frac{dx}{dt}(t) = v_1(t) = \frac{dg}{dt}(t)e_1$. Since the expression for the covariant derivative $\frac{D_x v}{dt}(t)$ on the hyperboloid is similar to that for the sphere with the Euclidean inner product replaced by the Lorentzian inner product, it follows that the derivatives $\frac{dg}{dt}e_i$ are given by the same expressions as in the spherical case except for $\frac{dg}{dt}e_2$ which is given by

$$\frac{dg}{dt}(t)e_2 = g(t)e_1 + \sum_{j=1}^{j=n} U_{2j}(t)g(t)e_{j+1}.$$

Therefore Darboux curves coincide with the solutions $g(t)$ of the following differential system:

(6)
$$\frac{dg(t)}{dt} = g(t)\begin{pmatrix} 0 & 1\ldots & & 0 \\ 1 & & & \\ \vdots & & U(t) & \\ 0 & & & \end{pmatrix}$$

with $U(t)$ arbitrary $so_n(\mathbb{R})$-valued, bounded and measurable functions on $[0,T]$.

Equation (4) that describes the Euclidean Darboux curves can be written in terms of coordinates on SE_n that allow for easy comparisons with non-Euclidean Darboux curves. These coordinates are obtained as follows:

The reference frame $e_1,\ldots,e_n$ when considered as an orthonormal basis for $\mathbb{E}^n$ identifies each point x of $\mathbb{E}^n$ with a column vector of coordinates $\begin{pmatrix} x_1 \\ \vdots \\ x_n \end{pmatrix}$ in $\mathbb{R}^n$, and identifies each $R \in SO_n$ with the matrix (R_{ij}) via $Re_i = \sum_{j=1}^{n} R_{ij}e_j$.

The correspondence between elements (x,R) of SE_n and column vectors in $\mathbb{R}^n$ and matrices in $SO_n(R)$ defines an isomorphism between $\mathbb{R}^n \rtimes SO_n(\mathbb{R})$ and SE_n. This isomorphism provides for an embedding of SE_n into $GL_{n+1}(\mathbb{R})$ by identifying $(x,R) \in \mathbb{R}^n \rtimes SO_n(\mathbb{R})$ with a matrix $g = \begin{pmatrix} 1 & 0 \\ x & R \end{pmatrix}$ in $GL_{n+1}(\mathbb{R})$.

Having made these notational adjustments, then differential system (4) may be written as

(7)
$$\frac{dg}{dt}(t) = g(t)\begin{pmatrix} 0 & 0 & \cdots & 0 \\ 1 & & & \\ \vdots & & U(t) & \\ 0 & & & \end{pmatrix}.$$

Equations $(5),(6)$ and (7) can now be described by a single parameter ϵ as

(8)
$$\frac{dg}{dt}(t) = g(t)\begin{pmatrix} 0 & -\epsilon & 0 & \cdots & 0 \\ 1 & & & & \\ \vdots & & & U(t) & \\ 0 & & & & \end{pmatrix}.$$

with the understanding that $\epsilon = 0, 1, -1$ and that $g(t)$ belongs to the group that corresponds to ϵ.

To focus on further similarities between these three cases it will be convenient to write G_ϵ for $\mathbb{R}^n \rtimes SO_n(\mathbb{R})$ when $\epsilon = 0$, SO_{n+1} when $\epsilon = 1$, and $SO(1, n)$ when $\epsilon = -1$. Then $M_n(\epsilon)$ will denote the underlying Riemannian space of G_ϵ, and π_ϵ will denote the projection from G_ϵ on $M_n(\epsilon)$.

The elastic problem of Kirchhoff is defined by an additional metric $\langle\langle U, V \rangle\rangle_g$ on the space of frame deformations $\frac{dg}{dt}$ which will be assumed left invariant in this paper. Then, Kirchhoff's elastic problem in $M_n(\epsilon)$ can be reformulated as the problem of finding the solution curve of equation (8) in the interval $[0, T]$ that minimizes the integral

(K)
$$\frac{1}{2} \int_0^T \langle\langle U(t), U(t) \rangle\rangle \, dt$$

relative to all Darboux curves $g(t)$ in G_ϵ that are generated by bounded and measurable matrices $U(t)$ in the interval $[0, T]$ and satisfy the given boundary conditions $g(0) = g_0$ and $g(T) = g_1$.

In the elastic problem of Euler, however, the matrices $U(t)$ in (8) are restricted to

(9)
$$U(t) = \begin{pmatrix} 0 & -u_1(t) & \cdots & -u_n(t) \\ u_1(t) & 0 & \cdots & 0 \\ \vdots & 0 & \cdots & 0 \\ u_n(t) & 0 & \cdots & 0 \end{pmatrix}.$$

and the curvature $\kappa(t)$ of the projected curve $x(t) = \pi_\epsilon(g(t))$ is given by

$$\kappa^2(t) = \sum_{i=1}^n u_i^2(t).$$

It follows that the elastic problem of Euler can be reformulated as an optimal problem of minimizing the integral

(E)
$$\frac{1}{2} \int_0^T \sum_{i=1}^n u_i^2(t) \, dt$$

over all solutions $g(t)$ of (7) generated by bounded and measurable matrices $U(t)$ described by (8) in the interval $[0, T]$ that satisfy, either the fixed boundary conditions $g(0) = g_0, g(T) = g_1$ or belong to fixed initial and terminal manifolds S_0, S_1 in G_ϵ.

THEOREM 1. *Assume that there exists a Darboux curve $g(t)$ defined on $[0, T]$ that satisfies the given boundary conditions for either of the preceeding optimal control problems. Then there is an optimal solution. That is, both elastic problems are well posed.*

PROOF. Since the proof is conceptually the same for both problems only the proof for Kirchhoff's elastic problem will be presented. Let g_0 and g_1 denote the given boundary conditions, and let $\mathrm{Traj}(g_0, g_1)$ denote the set of all solution curves $g(t)$ of (7) that are defined on the interval $[0, T]$ and conform to $g(0) = g_0$ and $g(T) = g_1$. Then $\mathrm{Traj}(g_0, g_1)$ is not empty by the assumption of the theorem.

Let β denote the infimum of $\int_0^T \big(U(t), U(t)\big) dt$ over all bounded and measurable matrices $U(t)$ in the interval $[0, T]$ that generate curves $g(t)$ in $\mathrm{Traj}(g_0, g_1)$, and let $g_n(t)$ denote a sequence of Darboux curves in $\mathrm{Traj}(g_0, g_1)$ generated by $U_n(t)$ such that

$$\beta = \lim \frac{1}{2} \int_0^T \langle\langle U_n(t), U_n(t)\rangle\rangle \, dt.$$

Then, $\{U_n\}$ is a bounded set in the Hilbert space $L^2([0, T])$ consisting of all measurable functions $U(t)$ that take values in the space of skew-symmetric matrices such that $\int_0^T \langle\langle U(t), U(t)\rangle\rangle \, dt < \infty$.

As is well known, any bounded set in a Hilbert space is weakly compact, and therefore $\{U_n\}$ contains a weakly convergent subsequence. For simplicity of notation it will be assumed that $\{U_n\}$ itself is weakly convergent. But then it is known in the literature on control theory that the trajectories $\{g_n(t)\}$ that originate at a fixed initial point g_0 converge uniformly whenever the corresponding controls $\{U_n\}$ converge weakly (for example, [**Ju2**], p.118).

Let U_∞ denote the weak limit of the sequence $\{U_n\}$, and let g_∞ denote the corresponding trajectory that satisfies $g_\infty(0) = g_0$. Since $\lim g_n = g_\infty$ is uniform, $g_\infty(T) = g_1$ and hence $g_\infty \in \mathrm{Traj}(g_0, g_1)$. It now follows that g_∞ is optimal because

$$\int_0^T \langle\langle U_\infty, U_\infty\rangle\rangle, dt = \lim_{n\to\infty} \int_0^T \langle\langle U_\infty, U_n\rangle\rangle \, dt$$

and

$$\int_0^T \langle\langle U_\infty, U_n\rangle\rangle \, dt \leq \left(\int_0^T \langle\langle\langle U_\infty, U_\infty\rangle\rangle, dt\right)^{1/2} \left(\int_0^T (U_n, U_n) \, dt\right)^{1/2}.$$

Therefore,

$$\left(\int_0^T \langle\langle U_\infty, U_\infty \rangle\rangle \, dt \right) \le \left(\int_0^T \langle\langle U_\infty, U_\infty \rangle\rangle \, dt \right)^{1/2} \lim \left(\int_0^T \langle\langle U_n, U_n \rangle\rangle \, dt \right)^{1/2}$$

$$= \left(\int_0^T \langle\langle U_\infty, U_\infty \rangle\rangle \, dt \right)^{1/2} \beta^{1/2}$$

and consequently,

$$\int_0^T \langle\langle U_\infty, U_\infty \rangle\rangle dt \le \beta \ .$$

Thus, g_∞ is optimal relative to the boundary conditions g_0 and g_1.

It is easy to prove that for each pair of boundary conditions there is an interval $[0, T]$ and a trajectory $g(t)$ of (8) that satisfies these boundary conditions. Since T corresponds to the length of the projected curve $x(t)$ the above statement could be paraphrased by saying that any boundary conditions can be matched by a curve that is sufficiently long. Therefore for any pair of boundary conditions there is an interval $[0, T]$ in which there is an optimal solution. In addition any curve in the base manifold can be lifted to a Darboux curve according to the next proposition.

THEOREM 2. *Any absolutely continuous curve $x(t)$ in $M_n(\epsilon)$ that satisfies $\| \frac{dx}{dt}(t) \| = 1$ in an interval $[0, T]$ is the projection of a Darboux curve with $U(t)$ given by (8).*

PROOF. It will be convenient to identify points x of $\mathbb{R}^n$ with the affine hyperplane $\begin{pmatrix} 1 \\ x \end{pmatrix}$ so that the projections of Darboux curves $g(t)$ are given by $\pi_\epsilon(g(t)) = g(t)e_1$ for each value of ϵ. Denote by K the subgroup of G_ϵ such that $Ke_1 = e_1$. It follows that $K = \{1\} \times SO_n(\mathbb{R})$. Let $g(t)$ denote any Darboux curve that projects onto the given curve $x(t)$ in $M_n(\epsilon)$. Then

$$g(t)e_1 = x(t) \text{ and } \frac{dx}{dt}(t) = v_1(t) = g(t)e_2.$$

.

If $g(t)$ is generated by $U(t)$ then let $h(t)$ denote the solution in $SO_n(\mathbb{R})$ of the equation

$$\frac{dh}{dt}(t) = h(t)U(t)$$

that satisfies $h(0) = I$. The matrix $U(t)$ can be written as $U(t) = V(t) + W(t)$ with

$$V = \begin{pmatrix} 0 & -U_{12} & \cdots & -U_{1n} \\ U_{12} & 0 & \cdots & 0 \\ \vdots & \vdots & 0 & \\ U_{1n} & 0 & \cdots & 0 \end{pmatrix} \text{ and } W = \begin{pmatrix} 0 & 0 & \cdots & & 0 \\ 0 & 0 & -U_{23} & \cdots & -U_{2n} \\ 0 & U_{23} & \cdots & & -U_{3n} \\ \vdots & \vdots & \vdots & & \vdots \\ 0 & U_{2n} & & \cdots & 0 \end{pmatrix}$$

If $w(t)$ is the solution of $\frac{dw}{dt}(t) = w(t)W(t)$ with $w(0) = I$ then the matrix $h_0(t) = h(t)w^{-1}(t)$ is the solution of $\frac{dh_0}{dt}(t) = h_0(t)U_0(t)$ where the matrix $U_0(t)$ is as in (8). Let $g_0(t) = \{1\} \times h_0(t)$ denote the embedding of h_0 in K. Then $g(t)g_0(t)$ is the desired Darboux curve and the proof is finished.

II. Cartan decomposition and the Generalized elastic problems

It will be useful to continue with the notations introduced in the preceeding demonstration whereby the Euclidean space $\mathbb{E}^n$ is identified with the affine hyperplane $\{1\} \times \mathbb{R}^n$ and where $K = \{1\} \times SO_n(\mathbb{R})$ denotes the isotropy subgroup of G_ϵ. Then the projection map $\pi_\epsilon : G_\epsilon \to M_n(\epsilon)$ is given by $\pi_\epsilon(g) = ge_1$. The isotropy group K acts on G_ϵ and satisfies $\pi(gK) = \pi(g)$. Therefore $M_n(\epsilon) = G_\epsilon/K$, i.e., G_ϵ is a principal K bundle over $M_n(\epsilon)$ ([**St**]).

Throughout the paper $\mathfrak{g}_\epsilon$ will denote the Lie algebra of G_ϵ and $\mathfrak{k}$ will denote the Lie algebra of K. Then $\mathfrak{p}_\epsilon$ denotes the complementary vector space consisting of all matrices $\begin{pmatrix} 0 & -\epsilon\alpha_1 - \cdots - \epsilon\alpha_n \\ \alpha_1 & \\ \vdots & 0 \\ \alpha_n & \end{pmatrix}$ for

$\alpha = \begin{pmatrix} \alpha_1 \\ \vdots \\ \alpha_n \end{pmatrix}$ in $\mathbb{R}^n$.

It is easy to verify that $\mathfrak{p}_\epsilon$ and $\mathfrak{k}$ satisfy the usual Lie algebraic relations

(10) $$\mathfrak{g}_\epsilon = \mathfrak{p}_\epsilon \oplus \mathfrak{k}, \ [\mathfrak{p}_\epsilon, \mathfrak{k}] = \mathfrak{p}_\epsilon, \ [\mathfrak{p}_\epsilon, \mathfrak{p}_\epsilon] \subseteq \mathfrak{k}, \ [\mathfrak{k}, \mathfrak{k}] = \mathfrak{k}$$

Additionally $[\mathfrak{p}_\epsilon, \mathfrak{p}_\epsilon] = 0$ when $\epsilon = 0$, and otherwise $[\mathfrak{p}_\varepsilon, \mathfrak{p}_\varepsilon] = \mathfrak{k}$.

A decomposition of a Lie algebra into a subalgebra and a vector space that satisfies (10) is generally known as a Cartan decomposition ([**Hg**]).

DEFINITION 2.1. *The trace form $Tr(A, B)$ is equal to the trace of* $\frac{1}{2}(AB)$ *for any matrices A and B in* $\mathfrak{g}_\epsilon$.

DEFINITION 2.2. *The horizontal distribution $\mathcal{H}_\epsilon$ is the distribution consisting of all left-invariant vector fields $X(g)$ in G_ϵ that take values in $\mathfrak{p}_\epsilon$. Integral curves of $\mathcal{H}_\epsilon$ will be called horizontal.*

In the language of principal bundles, $\mathcal{H}_\epsilon$ is known as a connection ([**St**]).

PROPOSITION 1.

a. *Let $(\pi_\epsilon)_*$ denotes the tangent map of the projection map π_ϵ. Then,*

$$(\pi_\epsilon)_*(\mathcal{H}_\epsilon(g)) = T_{\pi(g)}M_n(\epsilon).$$

b. *The Riemannian length $\|\frac{dx}{dt}(t)\|^2$ of the projected curve $x(t) = \pi_\epsilon(g(t))$, $\epsilon \neq 0$ of a horizontal curve $g(t)$ that is a solution of $\frac{dg}{dt}(t) = g(t)A(t)$ is equal to $(-\epsilon Tr(A(t), A(t)))^{\frac{1}{2}}$.*

PROOF. Let $x(t)$ be any curve in $M_n(\epsilon)$, and let $g(t)$ be any curve in G_ϵ that projects onto $x(t)$. Then $\frac{dg}{dt}(t) = g(t)A(t)$ for some curve of matrices $A(t)$ in $\mathfrak{g}_\epsilon$. Let $A(t) = A_0(t) + A_1(t)$ denote the decomposition of $A(t)$ with $A_0(t) \in \mathfrak{p}_\epsilon$ and $A_1(t) \in \mathfrak{k}$. If $g_0(t)$ denotes the solution of $\frac{dg_0}{dt}(t) = g_0(t)A_1(t)$ with $g_0(0) = I$ then $h(t) = g(t)(g_0(t))^{-1}$ is a horizontal curve that projects onto $x(t)$. Hence the first statement of the proposition follows.

To prove the second statement assume that $g(t)$ is a horizontal curve that is a solution of $\frac{dg}{dt}(t) = g(t)A(t)$. Then

$$x(t) = \pi_\epsilon(g(t)) = g(t)e_1,$$

therefore

$$\|\frac{dx}{dt}(t)\|^2 = \|\frac{dg}{dt}e_1\|^2 = \|g(t)A(t)e_1\|^2 = \|A(t)e_1\|^2.$$

The matrix $A(t)$ is in $\mathfrak{p}_\epsilon$, and therefore

$$A(t) = \begin{pmatrix} 0 & -\epsilon\alpha_1(t) - \cdots - \epsilon\alpha_n(t) \\ \alpha_1(t) & \\ \vdots & 0 \\ \alpha_n(t) & \end{pmatrix} \quad \text{for some } \alpha(t) = \begin{pmatrix} \alpha_1(t) \\ \vdots \\ \alpha_n(t) \end{pmatrix}$$

in $\mathbb{R}^n$. Then $A(t)e_1 = \begin{pmatrix} 0 \\ \alpha_1(t) \\ \vdots \\ \alpha_n(t) \end{pmatrix}$, and therefore

$$\|A(t)e_1\|^2 = \sum_{i=1}^{n} \alpha_i^2(t) = -\epsilon Tr(A(t), A(t)).$$

In view of the preceeding proposition it is then natural to define the following quadratic forms $\langle A, B \rangle_\epsilon$ on $\mathfrak{g}_\epsilon$.

DEFINITION 2.3. *Let A and B denote matrices in $\mathfrak{g}$ with decompositions*

$$A = A_0 + A_1 \text{ and } B = B_0 + B_1$$

with A_0 and B_0 in $\mathfrak{p}_\epsilon$ and A_1 and B_1 in $\mathfrak{k}$. Then

   i. $\langle A, B \rangle_\epsilon = -\epsilon Tr(A, B)$ when $\epsilon \neq 0$, and

   ii. $\langle A, B \rangle_0 = \langle A_0 e_1, B_0 e_1 \rangle - Tr(A_1, B_1)$ where $\langle A_0 e_1, B_0 e_1 \rangle$ denotes the Euclidean inner product of $A_0 e_1$ and $B_0 e_1$.

It follows that

1. $\langle A, B \rangle_{-1} = Tr(A, B)$ and $\langle A, B \rangle_1 = -Tr(A, B)$.

2. The trace form is negative definite on each $so_n(\mathbb{R})$, and therefore $\langle A, B \rangle_1$ is a positive definite form on $\mathfrak{g}_1 = so_{n+1}$.

   The quadratic form $\langle A, B \rangle_{-1}$ is positive definite on $\mathfrak{p}_{-1}$, and negative definite on $\mathfrak{k}$, and therefore it is non-degenerate on $so(1, n)$.

3. For $\epsilon \neq 0$

$$\langle A, [B, C] \rangle_\epsilon = \langle [A, B], C \rangle_\epsilon$$

for any matrices A, B, C.

4. The Euclidean quadratic form $\langle A, B \rangle_0$ is positive definite on $\mathfrak{g}_0 = \mathbb{R}^n \rtimes so_n(\mathbb{R})$ but does not satisfy the above invariance property.

5. The vector space $\mathfrak{p}_\epsilon$ is orthogonal relative to the form $\langle \cdot, \cdot \rangle_\epsilon$ for each ϵ.

6. Matrices $E_1(\epsilon), \ldots E_n(\epsilon)$, defined by

$$E_{i-1}(\epsilon) = \begin{pmatrix} 0 & -\epsilon e_i^* \\ e_i & 0 \end{pmatrix}$$

where $e_1^*, \ldots, e_n^*$ denotes the basis of row vectors corresponding to the standard basis of column vectors $e_1, \ldots e_n$ in $\mathbb{R}^n$, form an orthonormal basis for $\mathfrak{p}_\epsilon$ relative to the above quadratic form on $\mathfrak{p}_\epsilon$.

In terms of the above notations then Darboux curves $g(t)$ are the solutions of the differential system in G_ϵ

(11)
$$\frac{dg}{dt}(t) = g(t)(E_1(\epsilon) + U(t)).$$

as the matrices $U(t)$ range over bounded and measurable $\mathfrak{k}$-valued curves. In the language of control theory, the left invariant vector field $g \to gE_1(\epsilon)$ is the drift vector field of an affine control system (11) where the control functions $U(t)$ take values in the subalgebra $\mathfrak{k}$.

In the elastic problems of Euler the matrix $U(t)$ is of the form $U(t) = \begin{pmatrix} 0 & -u^*(t) \\ u(t) & 0 \end{pmatrix}$ and

$$\kappa^2(t) = \sum_{i=1}^n u_i^2(t) = -Tr(U(t), U(t)).$$

The submanifold S of G_ϵ determined by the first order jets in $M_n(\epsilon)$ consist of all matrices $g \in G_\epsilon$ for which the first and the second columns are fixed. Therefore the elastic problems of Euler are concerned with the minimum value of the integral $-\frac{1}{2} \int_0^T Tr(U(t), U(t)) \, dt$ over the trajectories $g(t)$ of control system (11) that either originate at a fixed point g_0 or a fixed manifold S_0 when $t = 0$ and terminate at a fixed point g_1 or a fixed manifold S_1 when $t = T$. The elastic problem of Kirchhoff is phrased similarly except that the integral that corresponds to the elastic energy is defined by a separate positive definite quadratic form $\langle\langle U(t), U(t) \rangle\rangle$.

The above descriptions elucidate the "non-holonomic" nature of the elastic problems. Recall the notion of holonomic constraints from the theory of mechanical systems: constraints in an n-dimensional manifold are called holonomic if their linear span defines an $n - m$ dimensional integrable distribution of permissible directions. The integral manifolds of this distribution contain the permissible motions of the system.

For the elastic problems the positive cone $\{E_1(\epsilon) + U : U \in \mathfrak{k}\}$ of admissible directions defined by (11) is the analogue of the distribution spanned by the constraints, and the reachable sets of (11) are analogues of the integral manifolds. Apart from the fact that the set of admissible directions is not even a vector space, the cone $\{E_1(\epsilon) + U : U \in \mathfrak{k}\}$ is an $\dfrac{n(n-1)}{2}$ dimensional cone in the Lie algebra $\mathfrak{g}$ whose dimension is equal to $n + \dfrac{n(n-1)}{2}$. Yet, in spite of the gap in the dimensions, the reachable set of (11) from any initial point is equal to G_ϵ, hence is not confined to any submanifold of lower dimension in the ambient space. In this sense the constraints that define the elastic problems might be called "completely non-integrable".

As variational problems on Lie groups the elastic problems may be seen as particular cases of a more general class of variational problems on any Lie group G that admits a Cartan decomposition $\mathfrak{g} = \mathfrak{p} \oplus \mathfrak{k}$ of the Lie algebra $\mathfrak{g}$ of G. This class is defined by an element B_0 of $\mathfrak{p}$, and a a positive definite quadratic form $\langle\langle , \rangle\rangle$ on $\mathfrak{k}$. Then

DEFINITION 2.4. THE GENERALIZED ELASTIC PROBLEM. *Find the minimum of $\frac{1}{2}\int_0^T \langle\langle U(t), U(t) \, dt$ among all absolutely continuous curves $g(t)$ in G that satisfy the fixed boundary conditions $g(0) = g_0$ and $g(T) = g_1$, and are the solutions of*

(GE)
$$\frac{dg(t)}{dt} = g(t)\big(B_0 + U(t)\big)$$

for some bounded and measurable $\mathfrak{k}$-valued function $U(t)$ on an interval $[0, T]$.

Typically, Cartan decompositions are induced by involutive automorphisms σ on a Lie group G. Recall that an automorphism $\sigma \neq I$ is involutive if $\sigma^2 = I$. If G is a connected Lie group that admits an involutive automorphism σ, then H_σ denotes the subgroup of G consisting of elements fixed by σ and H_0 denotes the connected component of H_σ that contains the identity.

The pair (G, K) is called symmetric if K is a closed subgroup of G such that $H_0 \subseteq K \subseteq H_\sigma$. If in addition $Ad_G(K)$ is compact then (G, K) is called Riemannian symmetric pair. Then the coset space G/K admits a left invariant Riemannian metric with respect to which $M = G/K$ is symmetric. [**Hg**].

For each Riemannian symmetric pair (G, K) the Lie algebra $\mathfrak{g}$ splits into a direct sum $\mathfrak{p} \oplus \mathfrak{k}$ with $\mathfrak{k}$ and $\mathfrak{p}$ induced by the automorphism σ. Since $\sigma^2 = I$, it follows that the tangent map σ_* at the identity is an isomorphism of $\mathfrak{g}$. Then,

$$\mathfrak{k} = \{A \in \mathfrak{g} : \sigma_*(A) = A\} \quad \text{and}$$

$$\mathfrak{p} \{A \in \mathfrak{g} : \sigma_*(A) = -A\}$$

from which it follows that

$$[\mathfrak{p}, \mathfrak{k}] \subset \mathfrak{p} , \quad [\mathfrak{p}, \mathfrak{p}] \subset \mathfrak{k} \text{ and } [\mathfrak{k}, \mathfrak{k}] = \mathfrak{k} .$$

The automorphism σ_ϵ that defines the space forms $M_n(\epsilon)$ are of the form $\sigma_\epsilon(g) = DgD$ for $g \in G_\epsilon$ where D is a diagonal matrix with its diagonal entries $-1, 1, \ldots 1$.

Here are some other examples of classical symmetric Riemannian spaces:

(a) The space $\mathcal{P}_n$ of positive definite matrices of determinant equal to 1 is realized as the coset space $SL_n(\mathbb{R})/SO_n(\mathbb{R})$ through the automorphism $\sigma(g) = (g^T)^{-1}$ with g^T denoting the matrix transpose of g in $SL_n(\mathbb{R})$. Evidently, $\sigma^2 = I$ and $SO_n(\mathbb{R})$ is the subgroup of fixed points of σ.

(b) Let G be any closed self adjoint ($G = G^T$) subgroup of $SL_n(\mathbb{R})$ and let $K = G \cap SO_n(\mathbb{R})$. Then $M = G/K$ is a totally geodesic submanifold of the space $\mathcal{P}_n$ in (a). For instance, both $G = SO(p,q)$ and $G = S_p(n)$ are self-adjoint closed subgroups of $SL_n(\mathbb{R})$.

(c) Every oriented Grassmannian manifold $G(k,n)$ is a symmetric Riemannian space. Let $G = SO_n(\mathbb{R})$ and let $K = SO_k(\mathbb{R}) \times SO_{n-k}(\mathbb{R})$, and let D_k be the diagonal matrix with diagonal entries $\underbrace{1, 1, \ldots, 1}_{k}$, $\underbrace{-1, -1, \ldots, -1}_{n-k}$. The automorphism $\sigma(g) = D_k g D_k^{-1}$ is involutive and fixes the elements $\begin{pmatrix} A_1 & 0 \\ 0 & A_2 \end{pmatrix}$ of the form $A_1 \in SO_k(\mathbb{R})$ and $A_2 \in SO_{n-k}(\mathbb{R})$.

Alternatively every k dimensional linear subspace S of $\mathbb{R}^n$ can be identified with the orthogonal reflection R_s given by $R_s(x) = x$, $x \in S$, and $R_s(x) = -x$, $x \in S^\perp$. Then $G(k,n)$ can be made into a topological space by requiring that the map $S \to R_S$ be a homeomorphism, in which case $G(k,n)$ becomes a closed, hence compact, submanifold of $SO_n(\mathbb{R})$. It is easy to verify that $G = SO_n(\mathbb{R})$ acts transitively on the space of reflections R_S and that the isotropy group of any k dimensional subspace S is equal to $K = SO_k(\mathbb{R}) \times SO_{n-k}(\mathbb{R})$. Thus $G(k,n) = SO_n(\mathbb{R})/SO_k(\mathbb{R}) \times SO_{n-k}(\mathbb{R})$.

The simplest generalization of Kirchhoff's elastic problem, which will be of interest for the comparisons with the equations of mechanical tops, consists of replacing the drift vector $E_1(\epsilon)$ in (11) with an arbitrary non-zero element B_0 in $\mathfrak{p}_\epsilon$. It is an easy exercise in controllability theory based on Lie saturate techniques to show that the control system

$$(12) \qquad \frac{dg}{dt}(t) = g(t)(B_0 + U(t))$$

remains controllable ([**Ju2**]), and therefore the corresponding generalized elastic problem is well posed. The same is true for a general symmetric space G/K, provided that the drift B_0 is restricted to regular element in $\mathfrak{p}$ ([**Ju2**]).

The questions of controllability, as interesting as they are in their own right, are somewhat tangential to the main theme of the paper, namel the contribution of the elastic problems to the theory of integrable Hamiltonian systems. They will not be pursued further. Instead, the exposition will shift to the Maximum Principle as a transitional device for the study of Hamiltonian systems.

III. The Maximum Principle and the Hamiltonians

The version of the Maximum Principle that will be presented in this paper, although far from the most general formulation, provides an easy path to the main issues in the paper and spares the reader unnecessary technical details ([**Ju2**], [**Su**]). The Maximum Principle will be stated for systems described by the differential equation:

$$(13) \qquad \frac{dx}{dt}(t) = X_0(x(t)) + \sum_{i=1}^{m} u_i(t) X_i(x(t))$$

on a smooth manifold M where $X_0, \ldots, X_m$ are smooth vector fields on M, and where the control functions $u_1(t), \ldots, u_m(t))$ are assumed to be bounded and measurable on compact intervals $[0, T]$.

It will be assumed that the total cost $\int_0^T f_0(x(t), u(t))\, dt$ of an integral curve $x(t)$ of (13) generated by a control function $u(t) = (u_1(t), \ldots, u_m(t))$ in the interval $[0, T]$ is defined by a smooth function f_0 from $M \times \mathbb{R}^m$ into $\mathbb{R}$. It will be assumed further that the initial sub manifold S_0 and the terminal sub manifold S_1 of M are given which may, in the limiting case, consist of single points x_0 and x_1. It will be convenient to refer to each pair of curves $(x(t), u(t))$ as a trajectory of (13) whenever $x(t)$ is an integral curve of (13) generated by a control curve $u(t)$.

A trajectory $(\bar{x}(t), \bar{u}(t))$ is said to be optimal (relative to the boundary data S_0, S_1 and T) if $\bar{x}(0) \in S_0$, and $\bar{x}(T) \in S_1$, and if furthermore

(14)
$$\int_0^T f_0(\bar{x}(t), \bar{u}(t)) \, dt \le \int_0^T f_0(x(t), u(t)) \, dt$$

for any trajectory $(x(t), u(t))$ in $[0, T]$ such that $x(0) \in S_0$ and $x(t) \in S_1$.

The Maximum Principle is a necessary condition of optimality expressed most naturally in the language of symplectic geometry of the cotangent bundle T^*M of M. Recall that the cotangent bundle of any manifold M is endowed with a canonical symplectic form ω that associates to each function F on T^*M a vector field $\vec{F}$ defined through the relation

(15)
$$dF_\xi(v) = \omega_\xi(\vec{F}(\xi), v)$$

valid for each point $\xi \in T^*M$ and each tangent vector v in $T_\xi(T^*(M))$. In conformity with the terminology established by mathematical physics, $\vec{F}$ is called the Hamiltonian vector field of F and F is called a Hamiltonian function, or simply a Hamiltonian.

Each vector field X on M defines a function H_X on T^*M defined by $H_X(\xi) = \xi(X(x))$ for all $\xi \in T_x^*(M)$. The function H_X is linear in each cotangent space $T_x^*(M)$. Non-autonomous vector fields $X(x, t)$ define time varying functions on $T^*M \times \mathbb{R}$. In particular, each control function $u(t)$ in (13) defines a time varying vector field on M which in turn defines a time varying Hamiltonian

$$H(\xi, u(t)) = H_0(\xi) + \sum_{i=0}^m u_i(t) H_i(\xi)$$

with $H_i(\xi) = \xi(X_i(x))$ for each $\xi \in T_x^*(M)$, $i = 1, \ldots, m$. There is another Hamiltonian function $\mathcal{H}_\lambda$ that depends on the dynamics (13) and also on the cost functional f_0. It is given by

$$\mathcal{H}_\lambda(\xi, u(t)) = \lambda f_0(\pi(\xi), u(t)) + H_0(\xi) + \sum_{i=0}^m u_i(t) H_i(\xi)$$

where λ is a parameter that can be either equal to 0 or -1 and where π is the canonical projection from T^*M onto M.

The integral curves of the Hamiltonian vector field $\vec{\mathcal{H}}_\lambda(\xi, u(t))$ are absolutely continuous curves $\xi(t)$ that satisfy

(16)
$$\frac{d\xi}{dt}(t) = \vec{\mathcal{H}}_\lambda(\xi(t), u(t))$$

for almost all t in each interval $[0, T]$. The reader can readily verify that the projection $(\pi(\xi(t)), u(t))$ is a trajectory of (13) for each trajectory $(\xi(t), u(t))$ of (16).

THE MAXIMUM PRINCIPLE. *Suppose that $(\bar{x}(t), \bar{u}(t))$ is an optimal trajectory relative to the function f_0 and the boundary data S_0, S_1, T as described by (14). Then $(\bar{x}(t), \bar{u}(t))$ is the projection of a trajectory $(\bar{\xi}(t), \bar{u}(t))$ of (16) on the interval $[0, T]$ such that:*
(1^0) If $\lambda = 0$, then $\bar{\xi}(t) \neq 0$ for any $t \in [0, T]$.
(2^0) The time varying Hamiltonian $\mathcal{H}_\lambda\big(\bar{\xi}(t), \bar{u}(t)\big)$ satisfies the following maximality condition:

$$\mathcal{H}_\lambda\big(\bar{\xi}(t), \bar{u}(t)\big) \geq \mathcal{H}_\lambda\big(\bar{\xi}(t), u\big)$$

for any u in $\mathbb{R}^m$ and almost all t in $[0, T]$.
(3^0) Furthermore, $\bar{\xi}(t)$ satisfies the transversality conditions $\bar{\xi}(0)(v) = 0$ for all tangent vectors v in S_0 at $\bar{x}(0)$, and $\bar{\xi}(T)(v) = 0$ for all tangent vectors v in S_1 at $\bar{x}(0)$.

Trajectories $(\xi(t), u(t))$ of the Hamiltonian system (16) that satisfy conditions (1^0) and (2^0) are called extremal curves. An extremal curve is called *normal* or *regular* if $\lambda = -1$, and *abnormal* if $\lambda = 0$. The Maximum Principle then can be restated in terms of the extremal curves by saying that every optimal trajectory is necessarily the projection of an extremal curve.

Abnormal extremal curves do not exist for optimal control problems in which every absolutely continuous curve $x(t)$ is a solution of the control system for some bounded and measurable control $u(t)$. In particular, that is the case for problems of mechanics subject to holonomic constraints, and also for problems of Riemannian geometry. For problems of mechanics the Lagrange's Principle of Least Action adapts to control theoretic interpretations and the use of the Maximum Principle efficiently leads to the total energy $H = T + V$ as the appropriate Hamiltonian for mechanical systems with the potential energy V ([**Ju2**]).

The problems of Riemannian geometry are slightly more interesting in the sense that the functional $f_0 = \sqrt{\|\frac{dx}{dt}\|}$ is singular at 0. However, this difficulty is easily removed by either restricting the paths to regular curves, i.e., by restricting to curves that satisfy $\frac{dx}{dt} \neq 0$, in which case the Riemannian problem becomes a time optimal control problem [**LuS**]), or by removing the square root in the integrand in which case the length functional becomes an energy functional. In this case to compensate for the loss of independence relative to the parameterizations of curves, the resulting Hamiltonian needs to be restricted to the energy level 1([**Ju2**] [**LuS**]). The abnormal extremals are ruled out in both of these formulations of the Riemannian problem.

However, in control problems there are constraints, and generally the number of control functions is less than the dimension of the state manifold M. In such situations abnormal extremal curves do exist, and have to be taken into consideration, although it turns out often, particularly for problems having symmetries, that the abnormal extremals can be ignored, either because they do not project onto optimal trajectories, or that the optimal trajectories are the projections of both the abnormal and the normal extremal curves. Unfortunately, there are no general criteria that rule out the possibility that an optimal trajectory is the projection of an abnormal extremal only.

1. The Hamiltonians of generalized elastic problems.

Recall that generalized elastic problems are defined by a Riemannian symmetric pair G, K where K is a compact Lie Group. The control system is defined by an element B_0 in the Cartan space $\mathfrak{p}$, the controls $U(t)$ take values in the Lie algebra $\mathfrak{k}$ of K, and the functional to be minimized is defined by a left-invariant positive definite quadratic form $\langle\langle A, B\rangle\rangle$ on $\mathfrak{k}$. In the applications that follow K is compact and the Cartan-Killing form is negative definite on $\mathfrak{k}$. Recall that the Cartan-Killing form is a scalar multiple of the trace form on groups that are semi-simple, and in particular that it is equal to the trace form on $SO_n(\mathbb{R})$. Consistent with our earlier notations $\langle\,,\,\rangle$ will denote the negative of the Killing form on $\mathfrak{k}$.

Let $A_1, \ldots, A_m$ denote a basis in $\mathfrak{k}$ that is orthonormal relative to the Cartan-Killing form which simultaneously diagonalizes the elastic form

$\langle\langle\,,\,\rangle\rangle$. Denote by $\lambda_1,\ldots,\lambda_n$ the diagonal entries of $\langle\langle\,,\,\rangle\rangle$ i.e. the numbers given by $\langle\langle A_i, A_j\rangle\rangle = \lambda_i\delta_{ij}$, $i=1,\ldots,m$. Relative to this choice of basis, control system (GE) can be written as (13) with $X_0,\ldots,X_m$ the left-invariant vector fields

$$X_0(g) = gB_0, X_1(g) = gA_1, X_2(g) = gA_2,\ldots,X_m(g) = gA_m$$

with the cost functional f_0 equal to $f_0(x(t),u(t)) = \frac{1}{2}\sum_{i=1}^{m}\lambda_i u_i^2(t)$ for each trajectory $(x(t),u(t)$ of (13).

The maximality condition (2^0) of the Maximum Principle then implies that the controls $u(t)$ that generate the regular extremal curves $\xi(t)$ are of feedback form

$$u_i(t) = \lambda_i^{-1}H_i(\xi(t)).$$

for $i=1,\ldots,m$. Then it readily follows that the regular extremal curves are the integral curves of a single Hamiltonian function H on T^*G of the form

$$H(\xi) = H_0(\xi) + \frac{1}{2}\sum_{i=1}^{m}\lambda_i^{-1}H_i^2(\xi).$$

for $\xi \in T^*G$.

In contrast to the regular case, controls $u(t)$ that generate the abnormal extremals $\xi(t)$ are not explicitly determined by the Maximum Principle. Instead, the maximality condition (2^0) degenerates to the constraints

$$H_1(\xi(t)) = 0, H_2(\xi(t)) = 0\ldots, H_m(\xi(t)) = 0.$$

which are then resolved by further differentiations, as will be demonstrated later on in the paper.

2. Mechanical systems and their Hamiltonians.

When it comes to the equations of motion for mechanical systems, the Maximum Principle offers certain advantages over the classical methods based on the Euler-Lagrange equation. The importance of this observation is particularly clear in the case of the equations of motion for mechanical tops as already demonstrated in ([**Ju2**]). Still, there are further new insights that can be obtained at the crossroads between control theory and mechanics as will be demonstrated later on in the paper. To get to some of the finer points however, it is most natural to go the beginning and re-derive the appropriate equations through the methods of optimal control.

The pendulum.   Consider first the planar case where a bob of mass m is attached to the pendulum of length l that is free to swing in a Euclidean plane $\mathbb{E}^2$ under the gravitational force. Assume that the Euclidean plane $\mathbb{E}^2$ has its origin at the fulcrum of the pendulum, and that it is oriented by an orthonormal frame $\vec{e}_1$, $\vec{e}_2$ so that the gravitational force is given by $\vec{F} = mg\vec{e}_1$.

Let $\big(\vec{a}_1(t), \vec{a}_2(t)\big)$ denote the orthonormal frame attached to the pendulum, and so adapted that the position $q(t)$ of the free end is in the direction of $\vec{a}_1(t)$ i.e., that $q(t) = la_1(t)$. (Figure 1)

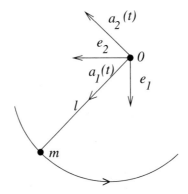

FIGURE 1

Let $R(t)$ denote the isometry defined by $R(t)\vec{e}_i = \vec{a}_i(t)$, $i = 1, 2$. Then each movement of the pendulum traces a curve $R(t)$ in $SO_2(\mathbb{R})$ given by $q(t) = l\vec{a}_1(t) = lR(t)\vec{e}_1$. The kinetic energy T associated with each movement $q(t)$ is given by $T = \dfrac{m}{2}\left\|\dfrac{dq(t)}{dt}\right\|^2 = \dfrac{1}{2}ml^2\left\|\dfrac{dR}{dt}\vec{e}_1\right\|^2$.

Since $R(t)$ is a curve in $SO_2(\mathbb{R})$, $\dfrac{dR(t)}{dt} = R(t)\begin{pmatrix} 0 & -u(t) \\ u(t) & 0 \end{pmatrix}$ for some function $u(t)$. Therefore, the kinetic energy is given by $T = \dfrac{1}{2}ml^2u^2(t)$.

The potential energy V in raising the pendulum from its equilibrium position $q(0) = \ell\vec{e}_1$ to an arbitrary position $q(t)$ is given by

$$V = -\int_0^t \vec{F} \cdot \frac{dq}{dt}dt = -mgl\int_0^t \left\langle \vec{e}_1, \frac{dR}{dt}\vec{e}_1 \right\rangle dt$$
$$= mgl\big(1 - \langle e_1, R(t)e_1 \rangle\big) ,$$

where $\langle \ , \ \rangle$ denotes the Euclidean metric on $\mathbb{E}^2$. Therefore the Lagrangian $L = T - V$ is given by $L(u, R) = \dfrac{1}{2}ml^2u^2 + mgl\langle e_1, R(t)e_1\rangle - mgl$.

According to the Principle of Least Action the movements of the pendulum minimize the integral $\int_{t_0}^{t_1} L\big(u(\tau), R(\tau)\big)d\tau$ over all paths $\dfrac{dR(t)}{dt} = u(t)X_1(R(t))$ subject to the appropriate boundary conditions, where X_1 denotes the left-invariant vector field $X_1(R) = R\begin{pmatrix} 0 & -1 \\ 1 & 0 \end{pmatrix}$.

The Maximum Principle immediately yields the Hamiltonian H on $T^*SO_2(\mathbb{R})$ of the form

(17)
$$H(\xi) = \frac{1}{2ml^2}H_1^2(\xi) - mgl\langle e_1, Re_1\rangle + mgl$$

for each ξ in $T^*SO_2(\mathbb{R})$ where H_1 denotes the Hamiltonian of the left-invariant vector field X_1. In addition, the Maximum Principle provides the relation between the angular velocity $u(t)$ and the angular momentum H_1 given by $u(t) = \dfrac{1}{ml^2}H_1(\xi(t))$. The equations of motion of the pendulum are the projections on $SO_2(\mathbb{R})$ of the integral curves $\xi(t)$ of the Hamiltonian vector field $\vec{H}$.

To recognize these equations in their familiar form parametrize $R(t)$ by an angle $\theta(t)$ defined by $R(t) = \begin{pmatrix} \cos\theta(t) & -\sin\theta(t) \\ \sin\theta(t) & \cos\theta(t) \end{pmatrix}$.

Then,
$$\frac{d\theta}{dt}(t) = u(t) = \frac{1}{ml^2}H_1(\xi(t)).$$

Since the Hamiltonian is constant along its integral curves it follows that
$$\frac{d}{dt}(H_1(\xi(t)) = -mgl\sin\theta(t),$$

and therefore,

(18)
$$\frac{d^2\theta}{dt^2}(t) + \frac{g}{l}\sin\theta(t) = 0$$

The spherical pendulum. The equations of motion for the spherical pendulum are obtained in a similar manner but with some modifications that will be explained below. Let e_1, e_2, e_3 denote a fixed orthonormal frame centered at the fulcrum of the pendulum and let $\vec{a}_1$, $\vec{a}_2$, $\vec{a}_3$ denote a moving frame also centered at the fulcrum of the pendulum and adapted to the pendulum so that the position of the free end q is given by $q = l\vec{a}_1$. Let R denote the rotation matrix defined by $\vec{a}_i = R\vec{e}_i, i = 1, 2, 3$.

This choice of polarization identifies S^2 as the quotient $SO_3(\mathbb{R})/K$ with K the isotropy subgroup of $SO_3(\mathbb{R})$ defined by $Ke_1 = e_1$. Evidently $K = \{1\} \times SO_2(\mathbb{R})$, hence K is isomorphic to $SO_2(\mathbb{R})$. Moreover, each R in $SO_3(\mathbb{R})$ can be regarded as an element of the orthonormal frame bundle of S^3 with Re_2, Re_3 considered an orthonormal frame at the point Re_1 of the sphere.

Then each path $R(t)$ in $SO_3(\mathbb{R})$ generates a path of the pendulum $q(t) = lR(t)e_1$. The velocity $\frac{dq}{dt}(t)$ is given by $\frac{dq}{dt}(t) = q(t) \times u(t)$ where $u(t) = (u_1(t), u_2(t), u_3(t))$ is defined by $\frac{dR}{dt}(t)$ through the formula

$$\frac{dR(t)}{dt} = \begin{pmatrix} 0 & -u_3(t) & u_2(t) \\ u_3(t) & 0 & -u_1(t) \\ -u_2(t) & u_1(t) & 0 \end{pmatrix} R(t) \ .$$

Therefore, vector $u(t)$ corresponds to the angular velocity $\omega(t)$ associated with the movement $q(t)$.

Since the pendulum does not rotate about its own axis the paths $R(t)$ in $SO_3(\mathbb{R})$ will be restricted by $u_1(t) = 0$. It is easy to show that each path $q(t)$ of the pendulum can be lifted to a path $R(t)$ in $SO_3(\mathbb{R})$ that is a solution curve of

$$\frac{dR(t)}{dt} = \begin{pmatrix} 0 & -u_3(t) & u_2(t) \\ u_3(t) & 0 & 0 \\ -u_2(t) & 0 & 0 \end{pmatrix} R(t)$$

for some functions $u_2(t)$ and $u_3(t)$.

It will be more consistent with the rest of the material in this paper to represent the preceeding equations in terms of left-invariant vector fields as follows. Replace $R(t), u(t)$ in the preceeding equation by $R^{-1}(t), -u(t)$. Then

(19)
$$\frac{dR}{dt}(t) = u_2(t)X_2(R(t)) + u_3(t)X_3(R(t))$$

where X_1, X_2, X_3 denote the left-invariant vector field on $SO_3(\mathbb{R})$ equal to the standard basis A_1, A_2, A_3 at the identity matrix I. The kinetic energy $T(t)$ associated with the movement $q(t)$ of the pendulum of mass m is given by

$$T(t) = \frac{1}{2}m \left\| \frac{dq}{dt} \right\|^2 (t) = \frac{1}{2}ml^2 \left\| R(t) \begin{pmatrix} 0 & -u_3(t) & u_2(t) \\ u_3 & 0 & 0 \\ -u_2(t) & 0 & 0 \end{pmatrix} e_1 \right\|^2$$

$$= \frac{1}{2}ml^2(u_2^2(t) + u_3^2(t)) \ ,$$

while the potential energy V retains the same form as in the planar case and is given by $V(t) = mgl\big(1 - \langle e_1, R(t)e_1 \rangle\big)$.

The Lagrangian $L = T - V$ lifts to a well defined function

$$L(R(t), u(t)) = \frac{1}{2}ml^2(u_2^2(t) + u_3^2(t)) - mgl\big(1 - \langle e_1, R(t)e_1 \rangle\big)$$

over the trajectories $(R(t), u(t))$ of (19). The Principle of Least Action is equivalent to the optimal control problem of minimizing $\int_0^T L(R(t), u(t))\, dt$ over the trajectories of (19) and the Maximum principle readily yields the Hamiltonian

(20) $$H(\xi) = \frac{1}{2ml^2}(H_2^2(\xi) + H_3^2(\xi)) - mgl\langle e_1, Re_1 \rangle + mgl,$$

where H_1 and H_2 are the momenta defined by $H_2(\xi) = \xi(R(A_2))$ and $H_3(\xi) = \xi(R(A_3))$ for ξ in $T_R(so_3^*(\mathbb{R}))$. Moreover, the extremal angular velocities $u_2(t)$ and $u_3(t)$ are related to the angular momenta H_2 and H_3 by

$$u_2(t) = \frac{1}{ml^2}H_2\big(\xi(t)\big) \text{ and } u_3(t) = \frac{1}{ml^2}H_3\big(\xi(t)\big).$$

A more detailed discussion of the corresponding Hamiltonian system $\frac{d\xi}{dt}(t) = \vec{H}(\xi(t))$ will be conducted in the next section.

The heavy tops. A rigid body in a three dimensional Euclidean space that is free to rotate around a fixed point under a constant gravitational force is known as a heavy top. The generalization to an arbitrary number of dimensions is due to T. Ratiu ([**Ra**]) and goes as follows.

Let $e_1, \ldots, e_n$ denote an absolute orthonormal frame, and let $a_1(t), \ldots, a_n(t)$ denote an orthonormal frame attached to the body with both frames centered at a fixed point O of the body. The absolute frame may be rotated so that the gravitational force is of the form

$$\vec{F} = -Ce_1 \text{ with } C \text{ the gravitational constant.}$$

Denote by $R(t)$ the isometry that relates the moving frame to the absolute frame. If $q = \begin{pmatrix} q_1 \\ \vdots \\ q_n \end{pmatrix}$ denotes the coordinate vector of a point $\vec{OP}$ in the absolute space, and if $Q = \begin{pmatrix} Q_1 \\ \vdots \\ Q_n \end{pmatrix}$ denotes the coordinate vector of the same point relative to the moving frame then

$$q = RQ .$$

With each movement of the body a point $q(t)$ undergoes velocity $\dfrac{dq}{dt} = \dfrac{dR}{dt}Q$ that generates the kinetic energy $\dfrac{1}{2}m\left\|\dfrac{dq}{dt}\right\|^2 =$ $\frac{1}{2}m\|V(t)R(t)Q\|^2$ where $V(t)$ denotes the antisymmetric matrix defined by $\dfrac{dR(t)}{dt} = V(t)R(t)$, and where $\|\cdot\|$ denotes the Euclidean norm. The total kinetic energy of the body is the aggregate of the contributions due to point masses and can be expressed as an n-dimensional integral in terms of the mass density ρ as follows:

$$T = \frac{1}{2}\int_{\text{Body}} \|V(t)R(t)Q\|^2\rho(Q)dQ = \frac{1}{2}\int_{\text{Body}} \|R^{-1}V(t)R(t)Q\|^2\rho(Q)dQ .$$

To simplify the preceding integrals let $U(t) = R^{-1}(t)V(t)R(t)$. Then $R(t)$ satisfies $\frac{dR^{-1}}{dt}(t) = R(t)U(t)$ and the associated kinetic energy is given by $\dfrac{1}{2}\displaystyle\int_{\text{Body}} \|U(t)Q\|^2\rho(Q)dQ.$

The kinetic energy may be expressed in terms of the quadratic form $\langle\langle U_1, U_2 \rangle\rangle$ in the Lie algebra $so_n(\mathbb{R})$ given by

$$\langle\langle U_1, U_2 \rangle\rangle = \int_{\text{Body}} \langle U_1Q, U_2Q \rangle \rho(Q)dQ$$

where $\langle U_1Q, U_2Q \rangle$ denotes the Euclidean product in $\mathbb{E}^n$. The quadratic form $\langle\langle\,,\,\rangle\rangle$ is positive definite and $T(U) = \dfrac{1}{2}\langle\langle U, U \rangle\rangle$. This formalism shows that the shape of each "rigid body" defines a left-invariant metric on $SO_n(\mathbb{R})$.

To find the expression for the potential energy it is convenient to introduce the center of mass Q_0 defined by $Q_0 \displaystyle\int_{\text{Body}} \rho(Q)dQ = \displaystyle\int_{\text{Body}} Q\rho(Q)dQ$. Then the potential energy is V is given by

$$V = C\int_0^t \int_{\text{Body}} \left\langle e_1, \frac{dR}{d\tau}Q \right\rangle \rho(Q)dQ\, d\tau = Cm\langle e_1, R(t)Q_0 \rangle ,$$

where $m = \int_{\text{Body}} \rho(Q)dQ$ stands for the total mass of the body.

As in the case of the pendulum, the Principle of Least Action is equivalent to the optimal control problem of minimizing the integral

$$\int_0^T (T - V)d\tau = \int_0^T (\frac{1}{2}\langle U(t), U(t) \rangle - Cm\langle R(t)e_1, Q_0 \rangle)\, dt$$

over all trajectories $(R(t), U(t))$ of $\dfrac{dR(t)}{dt} = R(t)U(t)$. Recall that R stands for the rotation that satisfies $Ra_i = e_1$ for $i = 1, \ldots, n$. In particular $R^{-1}e_1$ is the coordinate vector of e_1 relative to the moving frame.

The quadratic form $\langle\langle U, U \rangle\rangle$ induced by the kinetic energy can be expressed in terms of the trace form on $so_n(\mathbb{R})$ and a positive definite operator J, called the inertia tensor, such that $\langle\langle U, U \rangle\rangle = \langle U, JU \rangle$ for all U in $so_n(\mathbb{R})$.

Let $A_1, A_2, \ldots, A_m$ denote a basis in $so_n(\mathbb{R})$ that is orthonormal with respect to the trace form such that $\langle\langle A_i, A_j \rangle\rangle = \lambda_j \delta_{ij}$. The numbers $\lambda_1, \ldots, \lambda_n$ are the eigenvalues of the quadratic form $\langle\langle \, , \, \rangle\rangle$ and are called the principal moments of inertia in the literature on mechanics ([**Ar**]).

Curves $U(t)$ in $so_n(\mathbb{R})$ can be represented by m coordinate functions $u_1(t), \ldots, u_m(t)$ relative to the basis $A_1, \ldots, A_m$, and the optimal control problem of minimizing the integral

$$\int_0^T \left(\frac{1}{2} \sum_{i=1}^m \lambda_i u_i^2(t) + Cm\langle R^{-1}e_1, Q_0 \rangle \right) dt$$

over the trajectories of

$$\frac{dR}{dt}(t) = \sum_{i=1}^m u_i(t) X_i(R(t)),$$

with each X_i the left-invariant vector field on $SO_n(\mathbb{R})$ defined by $X_i(R) = RA_i$, produces the Hamiltonian H on the cotangent bundle of $SO_n(\mathbb{R})$ given by

(21)
$$H(\xi) = \frac{1}{2} \sum_{i=1}^m \frac{1}{\lambda_i} H_i^2(\xi) + Cm\langle R^{-1}e_1, Q_0 \rangle \ .$$

The extremal controls u_i are related to the momenta $H_i(\xi) = \xi(R(A_i))$ through the usual feedback formulas $\lambda_i u_i(\xi) == H_i(\xi)$ for each $i = 1, \ldots, m$. The Hamiltonian equations associated with H require additional theoretical ingredients which need to be developed first.

IV. The left-invariant symplectic form

The usual form of the Hamiltonian equations

(22)
$$\frac{dx}{dt} = \frac{\partial H}{\partial p}, \frac{dp}{dt} = -\frac{\partial H}{\partial x}$$

in canonical coordinates that one encounters in the literature on the calculus of variations is not the most suitable for variational problems on Lie groups that are either left or right invariant. To preserve these symmetries the cotangent bundle T^*G need be represented by $G \times \mathfrak{g}^*$ either via the right translations , or via the left translations depending which one is suitable for the problem at hand. Each choice of the trivialization of the cotangent bundle takes us outside the canonical coordinates when the group is non-Abelian.

In this paper the cotangent bundle will be trivialized by the left translations, which means that each point ξ in T^*G will be represented by a pair (g, ℓ) in $G \times \mathfrak{g}^*$ defined by the formula

$$\xi(g(A) = \ell(A)$$

for all $g \in G$ and all $A \in \mathfrak{g}$.

Remark. The correspondence $\xi \rightarrow (g, \ell)$ is formally defined by the left-translations $L_g(x) = gx$ as follows: The tangent map $(L_g)_*$ at the identity provides an isomorphism from $\mathfrak{g}$ onto the tangent space at g. It follows that the dual map $(L_g)^*$ maps T_g^*G onto $\mathfrak{g}^*$. Then the above correspondence is defined by

$$(L_g)^*(\xi) = \ell.$$

When G is a linear group, which will be the case for the groups in this paper, $(L_g)_*(A) = g(A)$. In such cases the notation is simplified, as already done in the earlier part of the paper, with $(L_g)_*(A)$ replaced by $g(A)$.

The symplectic formalism for real manifolds carries over to complex manifolds with essentially no alterations, and in particular applies to Lie groups. The extension to complex Lie groups is of importance as this paper will show. In anticipation of the material dealing with complex Lie groups, the symplectic structure of T^*G will be introduced simultaneously for either a real or a complex Lie group G.

Having identified T^*G with the product $G \times \mathfrak{g}^*$, the tangent bundle of T^*G is identified with the product $(G \times \mathfrak{g}^*) \times (\mathfrak{g} \times \mathfrak{g}^*)$ where it is understood that the second factor denotes the tangent vectors at the base point described by the first factor. Relative to this decomposition of the tangent bundle of T^*G, vector fields on T^*G will be written as pairs $\big(X(g, \ell), Y(g, \ell)\big)$ with $X(g, \ell) \in \mathfrak{g}$, $Y(g, \ell) \in \mathfrak{g}^*$, where (g, ℓ)

denotes the base point in $G \times \mathfrak{g}^*$. Then the canonical differential form θ on T^*G, known as the Liouville's form, is given by

$$\theta_{(g,\ell)}((X(g,\ell), Y(g,\ell)) = \ell(X(g,\ell)).$$

The exterior derivative $d\theta$ of θ is given by the Cartan form

(23) $$d\theta(V_1, V_2) = V_1(\theta(V_2)) - V_2(\theta(V_1)) + \theta([V_1, V_2])$$

To make use of the above formula tangent vectors need to be represented by vector fields. For that reason vectors in $T_{g,\ell}(G \times \mathfrak{g}^*)$ will be represented by pairs $(A, l) \in \mathfrak{g} \times \mathfrak{g}^*$ with the understanding that they correspond to the direction of the flow

$$\exp Vt(g, \ell) = (g \exp tA, \ell + tl)$$

at $t = 0$. Then,

$$V_2(\theta(V_1))(g, \ell) = \frac{d}{dt}(\theta_{\exp tV_2(g,\ell)}(V_1)|_{t=0} = \frac{d}{dt}(\ell + tl_2)(A_1)|_{t=0} = l_2(A_1).$$

The symplectic form ω is equal to $-d\theta$ and is given by

(24) $$\omega_{(g,\ell)}((A_1, l_1), (A_2, l_2)) = l_2(A_1) - l_1(A_2) - \ell([A_1, A_2]) .$$

The above form will be referred to as the left-invariant symplectic form on T^*G.

Remark. There is no established convention about the choice of signs in the above formula.

In addition to the choice of the sign, which is to a large degree arbitrary, the choice of the sign for the Lie bracket is also a matter of personal choice. So effectively, any choice of sign is possible in the above formula . The ambiguity in the sign is harmless as long as one is committed to a particular convention, although quoting different sources could cause problems since the conventions often are not explicitly stated. The convention in this paper is that the Lie bracket $[X, Y]$ of the left-invariant vector fields $X(g) = gA$ and $Y(g) = gB$ is given by $Z(g) = g[A, B]$ where $[A, B] = BA - AB$.

Consistent with the definitions used earlier in the paper, the Hamiltonian vector field $\vec{H}(g, \ell) = (A(g, \ell), l(g, \ell))$ of a function $H(g, \ell)$ on $G \times \mathfrak{g}^*$ is given by the formula

(25) $$\omega_{(g,\ell)}(A(g, \ell), l(g, \ell)), (B, m)) = dH_{(g,\ell)}(B, m)$$

for all tangent vectors (B, m) at (g, ℓ). Then,

$$dH_{g,\ell}(B, 0) = \frac{d}{dt} H(g \exp (Bt), \ell)|_{t=0} = \partial H_g(B)$$

and

$$dH_{g,\ell}(0, m) = \frac{d}{dt} H(g, \ell + tm))|_{t=0} = \partial H_l(m)$$

Since $\partial H_l(g, \ell)$ is a linear function on $\mathfrak{g}^*$ it is naturally identified with an element of $\mathfrak{g}$. It follows that

$$A(g, \ell) = \partial H_l(g, \ell) \text{ and } l(g, \ell) = -\partial H_g(g, \ell) - ad^* \partial H_l(g, \ell)(\ell),$$

and therefore the integral curves $(g(t), \ell(t))$ of $\vec{H}$ are the solution curves of

(26)
$$\frac{dg}{dt}(t) = g(t)\partial H_l(g(t), \ell(t)), \frac{d\ell}{dt}(t) = -\partial H_g(g(t), \ell(t)) - ad^* \partial H_l(g(t), \ell(t))(\ell(t))$$

where $ad^* A : \mathfrak{g}^* \to \mathfrak{g}^*$ is defined by $(ad^* A(\ell))(B) = \ell[A, B]$ for all $B \in \mathfrak{g}$.

DEFINITION 4.1. *The Poisson bracket* $\{H_1, H_2\}$ *of any functions* H_1 *and* H_2 *on* $G \times \mathfrak{g}^*$ *is a function on* $G \times \mathfrak{g}^*$ *defined by* $\{H_1, H_2\}(g, \ell) = \omega_{g,\ell}(\vec{H}_1(g, \ell), \vec{H}_2(g, \ell))$ *for all* (g, ℓ) *in* $G \times \mathfrak{g}^*$.

DEFINITION 4.2. *Functions* F *on* $G \times \mathfrak{g}^*$ *are said to be left-invariant if* $F(gh, \ell) = F(h, \ell)$ *for all* g *and* h *in* G *and all* ℓ *in* $\mathfrak{g}^*$.

It follows that the left-invariant functions on $G \times \mathfrak{g}^*$ are in exact correspondence with the functions on $\mathfrak{g}^*$. In particular, linear functions $H_A(\ell) = \ell(A)$ on $\mathfrak{g}^*$ defined by an element A in $\mathfrak{g}$ are left-invariant. Such functions are the Hamiltonian lifts of left-invariant vector fields on G, because $\xi(gA) = \ell(A)$ for any $\xi = (g, \ell)$ and any A in $\mathfrak{g}$. Moreover, if $H_1, \ldots, H_m$ is a collection of linear functions on $\mathfrak{g}^*$ generated by a basis $A_1, \ldots, A_m$ in $\mathfrak{g}$ then the vector $(H_1(\ell), \ldots, H_m(\ell))$ is the coordinate vector of ℓ relative to the dual basis $A_1^*, \ldots, A_m^*$. Consequently, any left invariant Hamiltonian H can be expressed as a function of the variables $H_1, \ldots, H_m$. The reader can easily verify that

(27) $$\{H_A, H_B\}(\ell) = \ell([A, B])$$

for any A and B in $\mathfrak{g}$.

It follows that the Hamiltonians associated with the generalized elastic problems are all left-invariant, while the only left-invariant Hamiltonian generated by a mechanical system discussed earlier occurs for the heavy top with the center of mass Q_0 situated at the fixed point of the body. Such a top is called the Euler top.

1. Coadjoint orbits and left-invariant Hamiltonians.

The integral curves $(g(t), \ell(t))$ of any left-invariant Hamiltonian vector field $\vec{F}$ are the solution curves of the following differential system

$$(28) \qquad \frac{dg}{dt} = g(t)dF_{\ell(t)} , \quad \text{and} \quad \frac{d\ell}{dt}(t) = -\mathrm{ad}^*(dF_{\ell(t)})(\ell(t))$$

as can be easily verified using equation (27). The Poisson bracket $\{F, H\}$ of left-invariant Hamiltonians F and H is left-invariant and is given by

$$(29) \qquad \{F, H\}(\ell) = \ell([dF_\ell, dH_\ell].$$

DEFINITION 4.3. *The coadjoint action of G on $\mathfrak{g}^*$ will be denoted by* Ad^*.

Recall that $Ad_g^*(\ell) = \ell \circ Ad_{g^{-1}}$ where $Ad_g(A) = gAg^{-1}$ for all $A \in \mathfrak{g}$.

PROPOSITION 4.1. *Let $(g(t), \ell(t)$ denote any solution of (28). Then $Ad_{g(t)}^*(\ell(t))$ is constant. That is, each solution $\ell(t)$ of (28) evolves on the coadjoint orbit of G through the initial point $\ell(0)$.*

PROOF. If $\frac{dg}{dt}(t) = g(t)dF_{\ell(t)}$ then $\frac{d}{dt}Ad_{g(t)} = -Ad_{g(t)} \circ ad(dF_{\ell(t)})$.

Therefore, $\ell(t) = Ad_{g^{-1}(t)}^*(\ell_0)$ for some ℓ_0 in $\mathfrak{g}^*$, as can be easily verified by differentiation. Hence $Ad_{g(t)}^*(\ell(t)) = \ell_0$.

PROPOSITION 4.2. *Each coadjoint orbit of G is a symplectic submanifold of $\mathfrak{g}^*$.*

PROOF. Let O_ℓ denote a coadjoint orbit $\{x : x = Ad_g^*(\ell), g \in G\}$. The tangent space $T_x O_\ell$ is equal to $\{ad^*A(x) : A \in \mathfrak{g}\}$ as can be easily verified by differentiating $y(t) = Ad_{x \exp tA}^*(x)$ at $t = 0$. Let

$$(30) \qquad B_x(ad^*A(x), ad^*B(x)) = x([A, B]).$$

Then B is a G invariant skew-symmetric 2-form on O_ℓ. To show that B is G invariant let $y = Ad_h^*(x)$. Then

$$Ad_h^* ad^* A(x) = Ad_h^*(x)adA = ad^* A(y).$$

Similarly $Ad_h^* ad^* B(x) = ad^* B(y)$, and therefore,

$$B_y\big(Ad_h^* ad^* A(x), Ad_h^* ad^* B(x)\big) = y([A, B]).$$

Since B is G invariant, it suffices to show that B is non-degenerate at a single point of O_ℓ. Let K denote the subgroup of G consisting of elements h such that $Ad_h^*(\ell) = \ell$, i.e., let K be the stabilizer of ℓ. Then O_ℓ can be regarded as the homogeneous space G/K. Let π denote the projection from G onto O_ℓ and let π_* denote its tangent map at the identity. Then

$$\pi_*(A) = \ell \circ adA.$$

It follows from above that $\pi_*(A) = 0$ if and only if A is in the Lie algebra of K. Hence, $\ell([A, B]) = 0$ for all $B \in \mathfrak{g}$ if and only if A belongs to the Lie algebra of K. This proves that B is non-degenerate on O_f.

The reader can easily verify that $\pi^*(B) = \omega$ where ω is the symplectic form given by (24). Therefore, B is closed and hence symplectic.

Hamiltonian equations on $\mathfrak{g}$. The passage from $\mathfrak{g}^*$ to $\mathfrak{g}$ is much simpler on semi-simple Lie groups due to the fact that the Cartan-Killing form $\langle\ ,\ \rangle$ is invariant and non-degenerate on $\mathfrak{g}$. The invariance of the form refers to the fact that

$$\langle [A, B], C \rangle = \langle A, [B, C] \rangle$$

for any matrices A, B, C in $\mathfrak{g}$.

The semi-simple Lie algebras. The Cartan-Killing form induces a natural identification of an element ℓ in $\mathfrak{g}^*$ with an element L in $\mathfrak{g}$ via the formula $\ell(A) = \langle L, A \rangle$ for all $A \in \mathfrak{g}$. In this identification solution curves $\ell(t)$ of (28) in $\mathfrak{g}^*$ are identified with curves $L(t)$ in $\mathfrak{g}$. It is an easy consequence of the invariance that $L(t)$ is the solution curve of

(31a)
$$\frac{dL(t)}{dt} = [dF_{\ell(t)}, L(t)] .$$

Together with

(31b)
$$\frac{dg(t)}{dt} = g(t) dF_{\ell(t)}$$

equations (31a) constitute an equivalent Hamiltonian system on $G \times \mathfrak{g}$.

Let us now return to the Hamiltonian $H = H_0(\ell) + \frac{1}{2} \sum_{i=1}^m \lambda_i^{-1} H_i^2(\ell)$ that corresponds to the generalized elastic problem (18). To begin with,

$$dH_{\ell(t)} = B_0 + \sum_{i=1}^m \lambda^{-1} H_i\big(\ell(t)\big) A_i .$$

The Cartan decomposition $\mathfrak{g} = \mathfrak{p} \oplus \mathfrak{k}$ induces a decomposition

$$L(t) = P(t) + M(t)$$

of any curve $L(t)$ in $\mathfrak{g}$ into the factors $P(t) \in \mathfrak{p}$ and $M(t) \in \mathfrak{k}$. Then equation

$$\frac{dL(t)}{dt} = [dH_{\ell(t)}, L(t)] = \left[B_0 + \sum_{i=1}^{m} \frac{H_i}{\lambda_i} A_i, L \right]$$

splits into the following pair of differential equations

(32a)
$$\frac{dM}{dt} = \big[\Omega(t), M(t) \big] + \big[B_0, P(t) \big] \ .$$

(32b)
$$\frac{dP}{dt} = \big[\Omega(t), P(t) \big] + \big[B_0, M(t) \big] \ .$$

where

$$\Omega(t) = \sum_{i=1}^{m} \frac{H_i(t)}{\lambda_i} A_i \ .$$

The equation for the abnormal extremal curves is similar in form to (32) except that $\Omega(t)$ is replaced by an open loop control $U(t)$ that needs to be determined so that the constraints $H_1(t) = \cdots = H_m(t) = 0$ are satisfied. But this means that $M(t) = 0$, and hence,

$$0 = \frac{dM}{dt} = \big[U(t), M(t) \big] + \big[B_0, P(t) \big] = \big[B_0, P(t) \big] \ ,$$

and

$$\frac{dP}{dt} = \big[U(t), P(t) \big] + \big[B_0, M(t) \big] = \big[U(t), P(t) \big] \ .$$

Thus abnormal extremal curves are non-zero solutions of

(33a)
$$\frac{dg(t)}{dt} = g(t) \big(U(t) \big) \ , \quad \frac{dP(t)}{dt} = \big[U(t), P(t) \big]$$

subject to the constraints

(33b)
$$\big[B_0, P(t) \big] = 0 \text{ and } M(t) = 0 \ .$$

The semi-direct product. Since the trace form is degenerate on the semi-direct products it cannot be used to set up a correspondence between $\mathfrak{g}^*$ and $\mathfrak{g}$. The quadratic form $\langle \, , \, \rangle_0$ given by Definition 2.3 is the next best alternative to the trace form and can be used instead.

Recall that the Cartan decomposition of the semi-direct product $\mathbb{R}^n \rtimes SO_n(\mathbb{R})$ is given by $\mathfrak{g} = \mathfrak{p} \oplus \mathfrak{k}$ with $\mathfrak{k}$ isomorphic to $so_n(\mathbb{R})$, and $\mathfrak{p}$ isomorphic with $\mathbb{R}^n$. Relative to this decomposition, matrices A in $\mathfrak{g}$ will be represented by the direct sums $A_0 + A_1$ with $A_0 \in \mathfrak{p}$ and $A_1 \in \mathfrak{k}$.

DEFINITION 4.4. *For each matrix A of the form*

$$A = \begin{pmatrix} 0 & 0 & \cdots & 0 \\ a_1 & 0 & \cdots & 0 \\ \vdots & \vdots & & \vdots \\ a_n & 0 & \cdots & 0 \end{pmatrix} \quad \hat{A} \text{ will denote the vector } \begin{pmatrix} a_1 \\ \vdots \\ a_n \end{pmatrix} \text{ in } \mathbb{R}^n.$$

The correspondence $A \to \hat{A}$ is exact and satisfies the following additional property:

$$\widehat{[A,B]} = -A\hat{B}$$

for any $A \in \mathfrak{k}$ and B in $\mathfrak{p}$.

In terms of the above notations, the quadratic form $\langle \ , \ \rangle_0$ of Definition 2.3 is given by

$$\langle A, B \rangle_0 = \hat{A}_0 \cdot \hat{B}_0 + \langle A_1, B_1 \rangle$$

where $\hat{A}_0 \cdot \hat{B}_0$ denotes the standard Euclidean inner product to differentiate it from the trace form $\langle A_1, B_1 \rangle = -\frac{1}{2}\operatorname{Tr}(A_1 B_1)$.

As in the semi-simple case each ℓ in $\mathfrak{g}^*$ will be identified with L in $\mathfrak{g}$ via the formula

$$\ell(A) = \hat{P} \cdot \hat{A}_0 + \langle M, A_1 \rangle \quad \text{for any} \quad A = A_0 + A_1 \quad \text{in} \quad \mathfrak{g} \ ,$$

and each L in $\mathfrak{g}$ will be written as $L = P + M$ with P in $\mathfrak{p}$ and M in $\mathfrak{k}$.

Suppose now that $\ell(t)$ is an extremal curve in $\mathfrak{g}^*$ corresponding to a Hamiltonian $\mathcal{H}$ on $\mathfrak{g}^*$. Then,

$$\frac{d\ell}{dt} = -(\operatorname{ad}^* d\mathcal{H}_{\ell(t)})\big(\ell(t)\big)$$

implies that

$$\left\langle \frac{dL}{dt}, A \right\rangle = \frac{d\ell}{dt}(A) = -\ell(t)\big([d\mathcal{H}, A]\big) = -\big\langle L(t), [d\mathcal{H}, A] \big\rangle$$

Since $d\mathcal{H}$ is in $\mathfrak{g}$, it can be decomposed as a direct sum $B + \Omega$ with $B \in \mathfrak{p}$ and $\Omega \in \mathfrak{k}$.

It follows that $[d\mathcal{H}, A] = [B+\Omega, A_0+A_1] = [B, A_1]+[\Omega, A_0]+[\Omega, A_1]$ and therefore

(34)
$$\frac{d\hat{P}}{dt} \cdot \hat{A}_0 + \left\langle \frac{dM}{dt}, A_1 \right\rangle = -\hat{P} \cdot (A_1\hat{B} - \Omega\hat{A}_0) - \big\langle M, [\Omega, A_1] \big\rangle$$

The preceding relations imply $\dfrac{d\hat{P}}{dt} \cdot \hat{A}_0 = \hat{P} \cdot (\Omega\hat{A}_0)$ for all A_0 in $\mathfrak{p}$, which in turn implies that $\dfrac{d\hat{P}}{dt} = -\Omega\hat{P}$, or that $\dfrac{dP}{dt} = [\Omega, P]$.

To obtain the equation for $M(t)$ the following lemma is needed

LEMMA 4.1. *Let a and b be any vectors in $\mathbb{R}^n$ and let A be any matrix in $so_n(\mathbb{R})$. Then*

$$Aa \cdot b = \langle a \wedge b, A \rangle$$

where $a \wedge b$ denotes the antisymmetric matrix that is defined by $(a \wedge b)(x) = (a \cdot x)b - (b \cdot x)a$ for all x in $\mathbb{R}^n$.

PROOF. The above is true for $a = e_i$ and $b = e_j$ and hence true for all vectors a and b.

It follows from (34) that

$$\left\langle \frac{dM}{dt}, A_1 \right\rangle = -\hat{P} \cdot A_1 \hat{B} - \langle M, [\Omega, A_1] \rangle$$
$$= -\langle \hat{B} \wedge \hat{P}, A_1 \rangle - \langle [M, \Omega], A_1 \rangle ,$$

and therefore

$$\frac{dM}{dt} = [\Omega, M] + \hat{P} \wedge \hat{B} .$$

The above can be stated as a proposition useful for future reference.

PROPOSITION 4.3. *The Hamiltonian equations of a left-invariant Hamiltonian $\mathcal{H}$ on the semi direct product $\mathfrak{p} \rtimes SO_n$ are given by*

(35) $$\frac{dM}{dt} = [\Omega, M] + \hat{P} \wedge \hat{B}, \, \frac{dP}{dt} = [\Omega, P], \frac{dg}{dt} = g(t)(d\mathcal{H})$$

where Ω and B denote the projections of $d\mathcal{H}$ on $\mathfrak{k}$ and $\mathfrak{p}$.

COROLLARY. *The regular extremal curves of the generalized elastic problem on $\mathbb{E}^n$ are the solution curves of equations (35) with $\Omega = \sum_{i=1}^{m} \frac{H_i}{\lambda_i} A_i$, and $B = B_0$.*
The abnormal extremals are the solution curves of

$$\frac{dg(t)}{dt} = g(t)(A_0 + U(t)) , \quad \frac{d\hat{P}}{dt} = -U(t)\hat{P}(t)$$

subject to the constraints $M(t) = 0$ and $\hat{P}(t) \wedge \hat{B}_0 = 0$.

Note that the Hamiltonian equations on the semi-direct product may be considered as a limiting case of the corresponding equations on the semi simple Lie algebras $\mathfrak{g}_\epsilon$ by the following arguments.

Instead of $\epsilon = \pm 1$ consider ϵ as a continuous non-negative parameter. Let $\mathfrak{g}_\epsilon$ denote the direct sum of $\mathfrak{p}_\epsilon$ and $\mathfrak{k}$ with $\mathfrak{p}_\epsilon$ the vector space of matrices of the form $\begin{pmatrix} 0 & -\epsilon a_1 - \cdots - \epsilon a_n \\ a_1 & \\ \vdots & 0 \\ a_n & \end{pmatrix}$ and $\mathfrak{k} = \{0\} \times so_n(\mathbb{R})$.

It is easy to check that $\mathfrak{g}_\epsilon$ is a Lie algebra for each real number ϵ. Evidently, $\mathfrak{g}_0 = \mathbb{R}^n \rtimes so_n(\mathbb{R})$ and $\mathfrak{g}_1 = so_{n+1}(\mathbb{R})$.

The trace form $\langle A, B \rangle_\epsilon = -\epsilon \frac{1}{2} \mathrm{Tr}(AB)$ is non-degenerate on $\mathfrak{g}_\epsilon$ for each non-zero value of parameter ϵ. Then equations (32) correspond to the Hamiltonian equations associated with a function $\mathcal{H}$ on $\mathfrak{g}_\epsilon^*$. Equations (35) follow from equations (32) by the following limiting procedure:

Let $E_1(\epsilon), \ldots, E_n(\epsilon)$ denote the standard basis for $\mathfrak{p}_\epsilon$. It is easy to verify that $\langle E_i(\epsilon), E_j(\epsilon) \rangle = \epsilon \delta_{ij}$. Let $p_1, \ldots, p_n$ denote the coordinates of a point p in $\mathfrak{p}_\epsilon$ relative to the dual basis $E_1^*(\epsilon), \ldots, E_n^*(\epsilon)$. For each p in $\mathfrak{p}_\epsilon$ let $P_\epsilon \in \mathfrak{p}_\epsilon$ be defined via the formula $p(A) = \langle P_\epsilon, A \rangle$ for all A in $\mathfrak{p}_\epsilon$. It follows that

$$P_\epsilon = \begin{pmatrix} 0 & -p_1 - \cdots - p_n \\ \frac{1}{\epsilon} p_1 & \\ \vdots & 0 \\ \frac{1}{\epsilon} p_n & \end{pmatrix}$$

.

If $\mathcal{H}$ is any function on $\mathfrak{g}_\epsilon^*$ then its Hamiltonian equations are given by

(36) $$\frac{dM}{dt} = [\Omega, M] + [B_\epsilon, P_\epsilon] , \quad \frac{dP_\epsilon}{dt} = [\Omega, P_\epsilon] + [B_\epsilon, M] .$$

where the differential $d\mathcal{H}$ is written as the sum $B_\epsilon + \Omega$ with $B_\epsilon = \sum_{i=1}^{n} \frac{\partial \mathcal{H}}{\partial p_i} E_i(\epsilon)$ and $\Omega = \sum_{i=1}^{m} \frac{\partial \mathcal{H}}{\partial H_i} A_i$.

Let $P(t) = \lim_{\epsilon \to 0} \epsilon P_\epsilon(t)$. Evidently $P(t) = \begin{pmatrix} 0 & 0 & \cdots & 0 \\ p_1 & 0 & \cdots & 0 \\ \vdots & \vdots & & \vdots \\ p_n & 0 & \vdots & 0 \end{pmatrix}$ belongs to $\mathfrak{g}_0$. Then equations (36) can be written as

(37) $$\frac{dM}{dt} = [\Omega, M] + \frac{1}{\epsilon}[B_\epsilon, \epsilon P_\epsilon], \text{ and } \epsilon \frac{dP_\epsilon}{dt} = [\Omega, \epsilon P_\epsilon] + \epsilon[B_\epsilon, M] .$$

It is easy to verify that $\frac{1}{\varepsilon}[B_\epsilon, P_\varepsilon] = \hat{B} \wedge \hat{P}$ with $\hat{B} = \begin{pmatrix} \frac{\partial \mathcal{H}}{\partial p_1} \\ \vdots \\ \frac{\partial \mathcal{H}}{\partial p_n} \end{pmatrix}$ and

$\hat{P} = \begin{pmatrix} p_1 \\ \vdots \\ p_n \end{pmatrix}$ and that $\lim_{\epsilon \to 0} \epsilon[B_\epsilon, M] = 0$. Therefore $\frac{dP}{dt} = \lim_{\epsilon \to 0}[\Omega, \epsilon P_\epsilon] =$

$[\Omega, P]$. Hence equations (35) follow from equations (37) as $\epsilon \to 0$.

2. Abnormal Extremals.

We shall restrict our considerations to the abnormal extremals generated by the elastic problems in the space forms $M_\epsilon = G_\epsilon/K$. Recall that the abnormal extremal curves are the solution curves of the following equations

$$\frac{dg}{dt}(t) = g(t)(B_0 + U(t)), \quad \frac{dP}{dt} = [U(t), P(t)])$$

subject to the constraints $M(t) = 0$ and $[B_0, P(t)] = 0$ for $\varepsilon \neq 0$, and $\hat{B}_0 \wedge \hat{P}$ for $\varepsilon = 0$.

The reader can easily check that $[B_0, P(t)]$ can be written as $\epsilon(\hat{B}_0 \wedge \hat{P})$,   where

$$B_0 = \begin{pmatrix} 0 & -\epsilon b_1 - \cdots - \epsilon b_n \\ b_1 & \\ \vdots & \quad 0 \\ b_n & \end{pmatrix} \text{ is represented by } \hat{B}_0 = \begin{pmatrix} b_1 \\ \vdots \\ b_n \end{pmatrix} \text{ and}$$

where

$$P(t) = \begin{pmatrix} 0 & -\epsilon p_1 - \cdots - \epsilon p_n \\ p_1(t) & \\ \vdots & \quad 0 \\ p_n(t) & \end{pmatrix} \text{ is represented by } \hat{P} = \begin{pmatrix} p_1 \\ \vdots \\ p_n \end{pmatrix}.$$

Hence all three cases are represented by $\hat{B}_0 \wedge \hat{P}(t) = 0$.

It is also easy to see that B_0 can be rotated to $\alpha E_1(\epsilon)$ for some real number α in which case $\hat{B}_0 = \alpha e_1$. In addition to this simplification, the following notations will be needed.

The sub algebra of $\mathfrak{k}$ consisting of the elements A such that $[E_1(\epsilon), A] = 0$ shall be denoted by $\mathfrak{k}_0$. Then K_0 will denote the Lie group generated by the exponentials in $\mathfrak{k}_0$. It follows that $\mathfrak{k}_0$ is isomorphic to $so_{n-1}(\mathbb{R})$, and that $K_0 = I \times SO_{n-1}(\mathbb{R})$ with $I = \{1\} \times \{1\}$.

THEOREM 4.1. *Let $g(t)$ and $P(t)$ be the solutions of*

$$\frac{dg}{dt}(t) = g(t)(E_1(\epsilon) + U(t)), \quad \frac{dP}{dt}(t) = [U(t), P(t)]$$

subject to the constraint $e_1 \wedge \hat{P}(t) = 0$. Then

(i) $P(t) = $ *constant and*

(ii) $g(t) = g(0)e^{tE_1(\epsilon)}h(t)$ *where $h(t)$ is an arbitrary curve in K_0 such that $h(0) = I$*

(iii) *If $g(t)$ is an optimal curve, then $h(t)$ in (ii)is an optimal curve relative to the left-invariant Riemannian metric in K_0 induced by the restriction of the elastic quadratic form $\langle\langle \ , \ \rangle\rangle$ to $\mathfrak{k}_0$.*

PROOF.

Solutions $P(t)$ of $\dfrac{dP}{dt} = [U(t), P(t)]$ satisfy $\langle P(t), P(t)\rangle = $ constant. Therefore, $\|\hat{P}(t)\| = $ constant.

Since $e_1 \wedge \hat{P}(t) = 0$, it follows that $p_2(t) = p_3(t) = \cdots = p_n(t) = 0$. Hence $P(t) = p_1 e_1$ for some non-zero constant p_1. This implies that $[U(t), P(t)] = 0$.

Since $P \neq 0$ it follows that the first row (and hence the first column of $U(t)$) is zero. Therefore, $E_1(\epsilon)$ and $U(t)$ commute. But then each solution of $\dfrac{dg}{dt} = g(\alpha E_1(\epsilon) + U(t))$ can be written as $g(t) = g(0)(\exp \alpha t E_1(\epsilon))h(t)$ where $h(t)$ is the solution of $\dfrac{dh(t)}{dt} = h(t)U(t)$ with $h(0) = I$. Since $U(t) \in \mathfrak{k}_0$ for all t, $h(t) \in K_0$ for all t. Thus (i) and (ii) are proved.

It is easy to see that if $g(t)$ is optimal in G then $h(t)$ must be optimal relative to the boundary points $h(0) = I$, $h(T) = e^{-\alpha E(\epsilon)T} \, g^{-1}(0)g(T)$ relative to the metric $\displaystyle\int_0^T \langle\langle U(t), U(t)\rangle\rangle dt)$ on $\mathfrak{k}_0$. $\square$

COROLLARY. *The projections $x(t) = g(t)e_1$ of abnormal extremal curves consist of parallel curves $x(t) = g(0)e^{E-1(\varepsilon)t}e_1$ with $x(0) = g(0)e_1$. Each such curve is also the projection of a regular extremal curve.*

PROOF. Since $K_0 \subset K$ and $Ke_1 = e_1$, the projection $x(t)$ of a curve $g(t) = g(0)e^{E_1(\epsilon)t}h(t)$ on M_ϵ is given by $x(t) = g(0)e^{E_1(\epsilon)t}e_1$. As $g(0)$ varies over G_ϵ these curves form parallel lines in the natural Riemannian metric on M_ϵ.

The curve $x(t)$ is also the projection of a regular extremal curve $g(t)$ that corresponds to $U(t) = 0$, $M = 0$, $P(t) = $ constant and $[E_1(\epsilon), P(t)] = 0$.

$\square$

On three dimensional space forms the extremal curves can be described very explicitly. The group K is isomorphic to $SO_2(\mathbb{R})$, and therefore all left invariant metrics on $SO_2(\mathbb{R})$ are scalar multiples of the canonical one. In particular, the geodesics $h(t)$ on K_0 generated by the left-invariant metric $\langle\langle\,,\,\rangle\rangle$ are of the form $h(t) = \exp tA$ for some matrix A in $\mathfrak{k}_0$. In particular, if $[E_1(\epsilon), U(t)] = 0$, then A is the rotation about the e_2 axis. Hence the abnormal optimal curves $g(t)$ are of the form $g(t) = g(0)e^{tE_1(\epsilon)}e^{u_1 A_1 t}$ for a constant u_1, where

$$E_1(\epsilon) = \begin{pmatrix} 0 & -\epsilon & 0 & 0 \\ 1 & 0 & 0 & 0 \\ 0 & 0 & 0 & 0 \\ 0 & 0 & 0 & 0 \end{pmatrix}, \text{ and where } A_1 = \begin{pmatrix} 0 & 0 & 0 & 0 \\ 1 & 0 & 0 & 0 \\ 0 & 0 & 0 & -1 \\ 0 & 0 & 1 & 0 \end{pmatrix}.$$

For $\epsilon = 0$, the curves $g(t)$ project onto parallel lines in $\mathbb{E}^3$ whose frames rotate with constant angular velocity in the plane perpendicular to e_2. For $\epsilon = 1$, the lines are replaced by circles, while for $\epsilon = -1$ the lines are replaced by hyperbolas.

In higher dimensions the study of abnormal extremal curves reduces to a study of left-invariant metrics on $SO_n(\mathbb{R})$.

2. The Hamiltonian systems of mechanical tops.

It is quite remarkable that mechanical tops, which are neither left nor right-invariant Hamiltonian systems on $T^*SO_n(\mathbb{R})$ form an invariant subsystem of the left-invariant Hamiltonian system associated with generalized elastic problem in $\mathbb{R}^n$. To demonstrate these phenomena it becomes necessary to express the Hamiltonian equations of various tops on the cotangent bundle of $SO_n(\mathbb{R})$, realized as the product of $SO_n(\mathbb{R}) \times so_n(\mathbb{R})^*$, and then show that the torque due to the gravitational force can be viewed as an element of the semi-direct product $\mathbb{R}^n \rtimes so_n(\mathbb{R})$. The basic idea is most simply illustrated through the Hamiltonian equations of the pendulum.

The pendulum and Elastic problem of Euler. The Hamiltonian associated with the pendulum is given by

$$H(\xi) = \frac{1}{2ml^2}H_1^2(\xi) - mgl\langle e_1, Re_1\rangle + mgl$$

as demonstrated by equation (18). The variable H_1 is the Hamiltonian lift of the left-invariant vector field $X_1(R) = RA_1$ with $A_1 = \begin{pmatrix} 0 & -1 \\ 1 & 0 \end{pmatrix}$. When $T^*SO_2(\mathbb{R})$ is realized as $SO_2(\mathbb{R}) \times so_2^*(\mathbb{R})$ the Hamiltonian H_1 and H become functions on $SO_2(\mathbb{R}) \times so_2^*(\mathbb{R})$ equal to $H_1(\ell) = \ell(A_1)$

and

$$H = \frac{1}{2ml^2} H_1^2(\ell) - mgl\langle e_1, Re_1\rangle + mgl.$$

The Hamiltonian equations of H are given by equations (26). To calculate $\partial H_R(A, \ell)$ let $R(t)$ denote the curve $R(t) = Re^{tA}$. Then

$$\partial H_R(A) = \frac{d}{dt} H(R(t), \ell)|_{t=0} = -mgl\langle R^*e_1, Ae_1\rangle.$$

and

$$\partial H_\ell = \frac{H_1(\ell)}{ml^2} A_1 = \begin{pmatrix} 0 & -\dfrac{H_1(\ell)}{ml^2} \\ \dfrac{H_1(\ell)}{ml^2} & 0 \end{pmatrix} = \Omega$$

Since $SO_2(\mathbb{R})$ is an Abelian group $ad^*(\partial H_\ell)(\ell) = 0$; therefore the integral curves of the Hamiltonian vector field $\vec{H}$ are given by

(38)
$$\frac{dR(t)}{dt} = R(t)\frac{H_1(\ell(t))}{ml^2} A_1 , \quad \frac{d\ell}{dt}(A) = mgl\langle e_1, RAe_1\rangle$$

for any A in $so_2(\mathbb{R})$.

To recognize these equations in their familiar form parametrize $R(t)$ by an angle $\theta(t)$ defined by $R(t) = \begin{pmatrix} \cos\theta(t) & -\sin\theta(t) \\ \sin\theta(t) & \cos\theta(t) \end{pmatrix}$. Then,

$$\frac{dR}{dt} = R(t)\frac{H_1(\ell(t))}{ml^2} A_1 \quad \text{reduces to} \quad \frac{d\theta}{dt} = \frac{H_1(\ell(t))}{ml^2} ,$$

and

$$\frac{dH_1(\ell(t))}{dt} = \frac{d\ell}{dt}(A_1) = mgl\langle e_1, R(t)A_1 e_1\rangle = -mgl\sin\theta(t) .$$

Hence,

$$\frac{d^2\theta}{dt^2}(t) = \frac{1}{ml^2}\frac{dH_1}{dt}(\ell(t) = -\frac{mgl}{ml^2}\sin\theta(t) ,$$

or

(39)
$$\frac{d^2\theta}{dt^2}(t) + \frac{g}{l}\sin\theta(t) = 0 .$$

While normally investigations of the planar pendulum end with equation (39), equation (38), which is hardly ever noticed in the literature on mechanics, shows connections to the semi-direct product and to the elastic problem of Euler as follows:

let $\hat{P}(t) = -mglR^{-1}(t)e_1$, and let $p_1(t)$, $p_2(t)$ denote its components. Then,

$$\frac{d\hat{P}}{dt} = -mgl\Omega(t)R^{-1}(t) = -\Omega(t)\hat{P}(t).$$

Identify $\ell \in so_2^*(\mathbb{R})$ with $M \in so_2(\mathbb{R})$ via the trace form. Then according to Lemma 4.1,

$$mgl\langle e_1, RAe_1\rangle = mgl\langle R^{-1}e_1 \wedge e_1, A\rangle = \hat{P}(t) \wedge e_1,$$

and therefore equation (38) can be written as

$$\frac{dR}{dt}(t) = R(t)\Omega(t)\,, \quad \frac{dM}{dt} = \hat{P}(t) \wedge e_1.$$

Together with $\frac{dP}{dt} = -\Omega(t)P(t)$ these equations are of the form

$$\frac{dM}{dt} = [\Omega, M] + \hat{P} \wedge \hat{B}\,, \quad \frac{dP}{dt} = [\Omega, P]\,, \quad \frac{dg}{dt} = g(t)(\partial H_\ell)$$

and can be regarded as the Hamiltonian equations of the left-invariant Hamiltonian

$$H = \frac{1}{2ml^2}\langle M, M\rangle + p_1 = \frac{1}{2ml^2}H_1^2 + p_1$$

on the semi-direct product $G = \mathbb{E}^2 \ltimes SO_2(\mathbb{R})$. Of course, the mathematical pendulum is confined to the "momentum" level $p_1^2 + p_2^2 = (mgl)^2$.

The spherical pendulum. The passage to the spherical pendulum requires only minor modifications. Recall the Hamiltonian for the spherical pendulum H given by equations (20):

$$H(\xi) = \frac{1}{2ml^2}(H_2^2(\xi) + H_3^2(\xi)) - mgl\langle e_1, Re_1\rangle + mgl$$

where H_2 and H_3 are the momenta defined by $H_2(\xi) = \xi(R(A_2))$ and $H_3(\xi) = \xi(R(A_3))$ for ξ in $T_R(so_3^*(\mathbb{R}))$.

In a manner analogous to the one used in the previous section, equations (26) can be transfered to $so_3(\mathbb{R})$ via the trace form. Then $\ell \in so_3(\mathbb{R})^*$ is associated with $M \in so_3(\mathbb{R})$, and the Hamiltonian equations of H are expressed in terms of the rotation matrix R and the momentum M as follows:

$$\frac{dR}{dt}(t) = R(t)\Omega(t)\,, \quad \text{and} \quad \frac{dM}{dt}(t) = [\Omega(t), M(t)] + \hat{P}(t) \wedge e_1$$

with

$$\Omega(t) = \begin{pmatrix} 0 & -\dfrac{H_3(t)}{ml^2} & \dfrac{H_2(t)}{ml^2} \\ \dfrac{H_3(t)}{ml^2} & 0 & 0 \\ -\dfrac{H_2(t)}{ml^2} & 0 & 0 \end{pmatrix}\,, M(t) = \begin{pmatrix} 0 & -H_3 & H_2 \\ H_3 & 0 & -H_1 \\ -H_2 & H_1 & 0 \end{pmatrix},$$

and $\hat{P}(t) = -mglR^{-1}(t)e_1$. Equivalently, the above can be written as

$$\frac{dR}{dt} = R(t)\Omega(t) \, , \quad \frac{dM}{dt} = \big[\Omega(t), M(t)\big] + \big[E_1, P(t)\big] \, .$$

under the identification of vectors $\hat{A} = \begin{pmatrix} a_1 \\ a_2 \\ a_3 \end{pmatrix}$ with antisymmetric ma-

trices $A = \begin{pmatrix} 0 & -a_3 & a_3 \\ a_3 & 0 & -a_1 \\ -a_2 & a_1 & 0 \end{pmatrix}$.

The extension to the semi-direct product is straightforward:

Each movement $q(l)$ of the pendulum defines a curve $g(t)$ in the group of motions where $g(t) = \begin{pmatrix} 1 & 0 \\ lq(t) & R(t) \end{pmatrix}$. Since $\hat{P}(t) = -mglR^{-1}e_1$,

$$\frac{d\hat{P}}{dt} = \hat{P}(t) \times \hat{\Omega}(t) \, , \quad \text{or} \quad \frac{dP}{dt} = \big[\Omega(t), P(t)\big] \, .$$

Together with

$$\frac{dM}{dt} = [\Omega, M] + [E_1, P]$$

these equations are Hamiltonian on the group of motions of $\mathbb{E}^3$ generated by

$$H = \frac{1}{2ml^2}(H_2^2(\xi) + H_3^2(\xi)) + p_1.$$

The above is the Hamiltonian (for the regular extremals) associated with Euler's elastic problem on $\mathbb{E}^3 \rtimes SO_3(\mathbb{R})$.

The heavy top. The equations for an n-dimensional heavy top follow the same pattern. The Hamiltonian H given by (21)

$$H(\xi) = \frac{1}{2}\sum_{i=1}^{m}\frac{1}{\lambda_i}H_i^2(\xi) + Cm\langle R^{-1}e_1, Q_0\rangle \, .$$

The extremal controls u_i are related to the momenta $H_i(\xi) = \xi(R(A_i))$ through the usual feedback formulas $\lambda_i u_i(\xi) = H_i(\xi)$ for each $i = 1, \ldots, m$. In the representation of the cotangent bundle of $SO_n(\mathbb{R})$ as the product $SO_n(\mathbb{R}) \times so_n(\mathbb{R})^*$ the dual variable ξ is replaced by the pair (R, ℓ) and the Hamiltonian equations are given by (26). It follows that

$$\partial H_\ell = \sum_{i=1}^{m}\lambda_i^{-1}H_i(\ell)A_i \text{ and } \partial H_R(A) = Cm\langle R^{-1}e_1, AQ_0\rangle.$$

As is customary in the literature on mechanics ∂H_ℓ will be denoted by Ω. Then the Hamiltonian equations (26) are given by

$$\frac{dR}{dt}(t) = R(t)\Omega(\ell(t))\,,\; \frac{d\ell}{dt}(A) = -Cm\langle R^{-1}e_1, AQ_0\rangle - ad^*(\Omega(\ell(t))(\ell(t))$$

Once more, linear functions ℓ will be identified with matrices M via the trace form in which case the preceeding equations become

$$(40) \qquad \frac{dR}{dt}(t) = R(t)\Omega(\ell(t))\,,\; \frac{dM}{dt}(t) = [\Omega(t), M(t)] + (CmR^{-1}e_1) \wedge Q_0.$$

To make the connection with equations (35) and the left-invariant Hamiltonians rename the variables $\hat{P}(t) = mCR^{-1}e_1$, $Q_0 = \hat{B}_0$, and identify the curves $x(t) = \int_0^t R(\tau)Q_0 d\tau$ and $R(t)$ with a curve $g(t) = \begin{pmatrix} 1 & 0 \\ x(t) & R(t) \end{pmatrix}$ in the group of motions $\mathbb{E}^n \ltimes SO_n(\mathbb{R})$. Then,

$$\frac{dg}{dt} = g(B_0 + \Omega(t)))\,,\; \frac{dM}{dt} = [\Omega(t), M(t)] + \hat{P} \wedge B_0,\; \text{and}\; \frac{dP}{dt} = [\Omega(t), P(t)]$$

where $P(t) = \begin{pmatrix} 0 & 0 & -\cdots- & 0 \\ p_1(t) & 0 & \cdots & 0 \\ \vdots & \vdots & & \vdots \\ p_n(t) & 0 & \cdots & 0 \end{pmatrix}$ and $B_0 = \begin{pmatrix} 0 & 0 & -\cdots- & 0 \\ b_1 & 0 & \cdots & 0 \\ \vdots & \vdots & & \vdots \\ b_n & 0 & \cdots & 0 \end{pmatrix}.$

These equations correspond to the left-invariant Hamiltonian H defined by

$$H = \frac{1}{2}\sum_{i=1}^m \frac{1}{\lambda_i}H_i^2 + \sum_{i=1}^n p_i b_i.$$

The three dimensional top is of major interest for applications and merits a few extra remarks on its relation to the corresponding elastic problem of Kirchhoff.

The top is defined by the coordinates Q_1, Q_2, Q_3 of the center of mass Q_0 relative to the orthonormal frame attached to the body, and by the principal moments of inertia $\lambda_1, \lambda_2, \lambda_3$. The principal moments of inertia correspond to the eigenvalues of the elastic quadratic form in the elastic problem of Kirchhoff, and the coordinates of the center of mass correspond to the way that the orthonormal frame is adapted to the tangent vector of the central line seen through the formula $x(t) = R(t)\hat{B}_0$. The variable $P(t)$, denoted usually by $\gamma(t)$ in literature on the top, denotes the torque due to the gravitational force. The torque is in the same direction as the position vector e_1 measured relative to the frame that is attached to the moving body.

The recognition of tops as invariant subsystems of the elastic problem re-orients our understandings of the classical theory of tops, and shifts our focus to the elastic problem and its geometric generalizations. This shift leads to new interpretations of old results and suggests exciting possibilities for further generalizations. To begin with, the embedding of the top into a left-invariant Hamiltonian system on a Lie group automatically implies certain symmetries of the system, as will be demonstrated in the next section.

V. Symmetries and the conservation laws

A Hamiltonian function H on a symplectic manifold M of dimension $2n$ is said to be *integrable* (or completely integrable) if there exist functions $\varphi_2, \ldots, \varphi_n$ on M that together with $\varphi_1 = H$ satisfy the following two properties:

(7) (i) $\varphi_1, \ldots, \varphi_n$ are functionally independent. Functional independence is understood in the local sense, that is, that the differentials

(8) $d\varphi_1, \ldots, d\varphi_n$ are linearly independent for an open (often dense) subset of M.

(9) (ii) The functions $\varphi_1, \ldots, \varphi_n$ are in involution, that is, they Poisson commute among each other.

Recall that the Poisson bracket $\{\varphi, \psi\}$ is a function defined by the symplectic form ω on M through the following equivalent conditions:

$$\{\varphi, \psi\}(x) = \omega_x\big(\vec{\varphi}(x), \vec{\psi}(x)\big) = \frac{d}{dt}(\varphi \cdot \exp t\vec{\psi})(x)\Big|_{t=0}$$

with $\vec{\varphi}$ and $\vec{\psi}$ denoting the Hamiltonian vector fields induced by the functions φ and ψ.

In the literature on mechanics, a function φ is called an integral of motion for the Hamiltonian H if $\{\varphi, H\} = 0$, that is, if φ is constant along the flow of $\vec{H}$. In this terminology then functions $\varphi_1, \ldots, \varphi_n$, defined by (ii) above are all integrals of motion for each other. The maximal number of functions $\varphi_1, \ldots, \varphi_m$ that are functionally independent and in involution is equal to $\frac{1}{2} \dim M$.

If H is an integrable system then each level set $\{x \,:\, \varphi_1(x) = c_1, \ldots, \varphi_n(x) = c_n\}$ is an n-dimensional submanifold of M. Such submanifolds are called Lagrangian. They are sub manifolds of M of maximal dimension on which the symplectic form vanishes. The connected component through each point x_0 of each level set $\{x \,:\, \varphi_i(x) = c_i \,,\, i \leq$

n} is equal to the orbit through x_0 of the commutative family of Hamiltonian vector fields $\{\vec{\varphi}_1, \ldots, \vec{\varphi}_n\}$.

The left-invariant Hamiltonians on a Lie group G always have extra integrals of motion due to symmetries induced by left-invariance.

DEFINITION 5.1. *Functions φ on $G \times \mathfrak{g}^*$ will be called conservation laws for G if φ is constant along the flow of any left-invariant Hamiltonian on G.*

Since the right invariant vector fields on G commute with left-invariant vector fields, the Hamiltonians of right-invariant vector fields Poisson commute with any left-invariant Hamiltonian H as explained in [**Ju2**]. Therefore, the Hamiltonian of any right-invariant vector field is a conservation law on G. The maximal number of such functions which Poisson commute with each other is determined by the rank of the Lie algebra $\mathfrak{g}$. Recall that the rank of $\mathfrak{g}$ is equal to the dimension of the maximal commutative sub algebra of $\mathfrak{g}$.

For instance, the rank of $SL_n(\mathbb{R})$ is $n-1$ equal to the dimension of diagonal matrices of trace zero. The rank of the semi-direct product is n, determined by n-commuting translations. The rank of $SO_4(\mathbb{R})$ and $SO_5(\mathbb{R})$ is two and so is the rank of $SO(1,3)$ and $SO(1,4)$ (see [**Hg**] for further details).

In addition to the Hamiltonians generated by the right-invariant vector fields there are other conservation laws, known as Casimir functions. For Hamiltonian systems on semi-simple Lie groups with the decomposition $\mathfrak{g} = k \oplus \mathfrak{p}$, the Hamiltonian equations (equation (32)) are of the form:

$$\frac{dM}{dt} = [dH, M] + [B, P], \quad \text{and} \quad \frac{dP}{dt} = [dH, P] + [B, M]$$

and $\varphi = \langle M, M \rangle + \langle P, P \rangle$ with $\langle \, , \, \rangle$ equal to the Killing form is an integral of motion for H.

The verification is simple:

$$\frac{1}{2}\frac{d}{dt}\langle M, M \rangle + \frac{1}{2}\frac{d}{dt}\langle P, P \rangle = \langle [dH, M] + [B, P], M \rangle + \langle [dH, P] + [B, M], P \rangle$$
$$= \langle dH, [M, M] \rangle + \rangle [B, P], M \rangle + \langle dH, [P, P] \rangle + \langle [B, M], P \rangle = 0 \, ,$$

and thus $\langle M(t), M(t) \rangle + \langle P(t), P(t) \rangle$ is constant along extremal curves.

This integral of motion can be also obtained by a more systematic procedure based on the left-invariant symmetries as follows:

the functions $\varphi(g, \ell) = \ell(g^{-1}Ag)$ are integrals of motion for any left-invariant Hamiltonian H, as was shown by Proposition 4.1. On semi-simple Lie groups, $\ell \in \mathfrak{g}^*$ is identified with L in $\mathfrak{g}$, and $\ell(g^{-1}Ag) = \langle L, g^{-1}Ag \rangle = \langle gLg^{-1}, A \rangle$. Therefore, $g(t)L(t)g^{-1}(t) = \text{constant}$ for any extremal curve $(g(t), L(t))$. Consequently, the trace of each power $L^j(t)$ must be a constant of motion, or equivalently, the coefficients of the characteristic polynomial of $L(t)$ must be constants of motion. In particular, the trace of L^2 is equal to $\langle M, M \rangle + \langle P, P \rangle$.

The traces of L^j for $j \geq 3$ provide additional integrals of motion as demonstrated below. Recall that on $\mathfrak{g}_\epsilon$ with $\epsilon \neq 0$ L is defined by

$$\ell(A) = \langle L, A \rangle_\epsilon \text{ for all } A \in \mathfrak{g}_\varepsilon .$$

where $\langle L, A \rangle_\epsilon$ is defined by Definition 2.3. Then $L = M + P$ implies that

$$\ell(A) = \langle M, A \rangle \ \text{ for } A \in \mathfrak{k}, \text{ and } \ell(A) = \langle P, A \rangle \text{ for } A \in \mathfrak{p}_\varepsilon.$$

Since $\mathfrak{k} \approx so_n(\mathbb{R})$ (independently of ϵ) it follows that $M = \sum_{i=1}^{m} M_i A_i$ for any orthonormal basis $A_1, \ldots, A_m$ in $\mathfrak{k}$ relative to $\langle \ , \ \rangle$ where $M_1, \ldots, M_m$ denote the coordinates of $\ell \in \mathfrak{k}^*$ in the dual basis $A_1^* \ldots, A_m^*$. Similarly P is represented by the matrix

$$P = \begin{pmatrix} 0 & -p_1 - \cdots - p_n \\ \frac{1}{\epsilon}p_1 & \\ \vdots & 0 \\ \frac{1}{\epsilon}p_n & \end{pmatrix}$$ where $p_1, \ldots, p_n$ denote the dual coor-

dinates of $p \in \mathfrak{p}_\epsilon^*$ relative to the standard basis $E_1(\epsilon), \ldots, E_n(\epsilon)$ in $\mathfrak{p}_\epsilon$.

Therefore L can be written simply as $L = \begin{pmatrix} 0 & -p^T \\ \frac{1}{\varepsilon}p & M \end{pmatrix}$ with M an antisymmetric $n \times n$ matrix, and p the column vector with entries $p_1, \ldots, p_n$.

In particular, $I_2 = \langle L(t), L(t) \rangle_\epsilon = \|p\|^2 + \epsilon \|M(t)\|^2$, and therefore I_2 is also a conservation law. The function I_2 may be regarded as a continuous function of ϵ which leads to an integral of motion even for $\epsilon = 0$, as can be easily verified through the equations (34).

Consider now the higher powers of L. It follows that

$$L^2 = \begin{pmatrix} -\frac{1}{\varepsilon}\|p\|^2 & -p^T M \\ \frac{1}{\varepsilon}Mp & M^2 - \frac{1}{\varepsilon}p \otimes p \end{pmatrix},$$

and therefore

$$-\epsilon \mathrm{Tr}(L^3) = p^* M p - \epsilon Tr(M^3) + \epsilon Tr(M p \otimes p).$$

Since M is antisymmetric, $p^* M p = 0$ and $Tr(M^3) = 0$. Moreover, $Tr(M(p \otimes p)) = p^* M p$. Hence, $Tr(L^3) = 0$.

An easy calculation shows that

$$\mathrm{Tr}(L^4) = \frac{1}{\varepsilon^2}\|p\|^4 - \frac{2}{\varepsilon} M p \cdot p^T M + \mathrm{Tr}(M^2 - \frac{1}{\varepsilon} p \otimes p)^2$$

where $x \cdot y$ denotes the Euclidean inner product in $\mathbb{R}^n$. Since the matrix M is antisymmetric, $M p \cdot p^T M = -\|M p\|^2$ and the above can be rewritten as

$$\mathrm{Tr}(L^4) = \frac{2}{\varepsilon^2}\|p\|^4 + \frac{4}{\varepsilon}\|M p\|^2 + \mathrm{Tr}(M^4) .$$

But then

$$I_\epsilon = \frac{1}{\epsilon}\left(I_2^2 - \frac{\epsilon^2}{2}\mathrm{Tr}(L^4)\right)$$
$$= \frac{1}{\epsilon}\left(\|p\|^4 + 2\epsilon\|M\|^2\|p\|^2 + \epsilon^2\|M\|^4\right) - \frac{1}{\epsilon}\left(\|p\|^4 + 2\epsilon\|M p\|^2 + \epsilon^2 \frac{1}{2}Tr(M)^4\right)$$
$$= \|M\|^2\|p\|^2 - \|M p\|^2 + \epsilon\left(\|M\|^4 - \frac{1}{2}Tr(M^4)\right)$$

is an integral of motion for any left-invariant Hamiltonian H on $\mathfrak{g}_\epsilon$ for $\epsilon \neq 0$. Again, I_ϵ can be regarded as a continuous function of ϵ which in the limit as $\epsilon \to 0$ is equal

$$I_0 = \|M\|^2\|p\|^2 - \|M p\|^2.$$

This limit is an integral of motion for the semi- direct product (equations (34)) as can be verified directly.

For $n = 4$, M is a 3×3 antisymmetric matrix $\begin{pmatrix} 0 & -M_3 & M_2 \\ M_3 & 0 & -M_1 \\ -M_2 & M_1 & 0 \end{pmatrix}$

Hence,

$$I_0 = \|M\|^2\|p\|^2 - \|M p\|^2 = \|M\|^2\|p\|^2 - \|\hat{M} \wedge p\|^2$$
$$= (\hat{M} \cdot p)^2 = (M_1 p_1 + M_2 p_2 + M_3 p_3)^2$$

Function $I_3 = M_1 p_1 + M_2 p_2 + M_3 p_3$ is a conservation law for any six-dimensional group G_ϵ for the following reasons. Any antisymmetric 3×3 matrix M can be expressed as $M = u \wedge v$ for some vectors u and v. Then $\|M\|^4 - \frac{1}{2}Tr M^4 = 0$, and the function $I_\epsilon = \|M\|^2\|p\|^2 - \|M p\|^2$

is an integral of motion valid for all ϵ. To explain further, suppose that $M = u \wedge v$ for some vectors u and v in $\mathbb{R}^n$. Then,

$$M^2 x = u\big((x \cdot v)(u \cdot v) - (x \cdot u)\|v\|^2\big) - v\big((x \cdot v)\|u\|^2 - (x \cdot u)(u \cdot v)\big)$$

Hence,

$$\langle M, M \rangle = -\frac{1}{2} \sum_{i=1}^{n} e_i \cdot M^2 e_i = \|u\|^2 \|v\|^2 - \langle u, v \rangle^2,$$

and

$$Tr(M^4) = \sum_{i=1}^{n} e \cdot M^4 e = \sum_{i=1}^{n} M^2 e_i \cdot M^2 e_i = \sum_{i=1}^{n} \|u(v_i \cdot u \cdot v - u_i \|v\|^2) - v(v_i \|u\|^2 - u_i(u \cdot v))\|^2$$

It follows that

$$
\begin{aligned}
Tr(M^4) &= \sum_{i=1}^{n} (\|u\|^2 (v_i(u.v) - u_i \|v\|^2)^2 + \|v\|^2 (v_i \|u\|^2 - u_i(u \cdot v))^2 \\
&\quad - 2(u \cdot v)(v_i(u \cdot v) - u_i \|v\|^2)(v_i \|u\|^2 - u_i(u \cdot v)) \\
&= 2(\|u\|^4 \|v\|^4 - 2\|u\|^2 \|v\|^2 (u \cdot v)^2 + (u \cdot v)^4) \\
&= 2\langle M, M \rangle^2,
\end{aligned}
$$

and therefore, $\frac{1}{2} Tr(M^4) - \|M\|^4 = 0$

The preceding developments imply that all elastic problems on two dimensional space forms are completely integrable since $I_1 = H$, $I_2 = p_1^2 + p_2^2 + \epsilon(M_1^2 + M_2^2)$, and the Hamiltonian I_3 of any right-invariant vector field form three functionally independent integrals of motion in involution with each other.

On three-dimensional space forms $I_1 = H$, $I_2 = (p_1^2 + p_2^2 + p_3^2) + \epsilon(M_1^2 + M_2^2 + M_3^2)$ and $I_3 = M_1 p_1 + M_2 p_2 + M_3 p_3$ are the integrals of motion for any left-invariant Hamiltonian H. Since the rank of $\mathfrak{g}_\epsilon$ for $\epsilon \neq 0$ is 2 there are two additional integrals of motion, namely the Hamiltonians of any two commuting right-invariant vector fields that are in involution with each other and in involution with I_1, I_2, I_3. The rank of the semi-direct product here is 3, but only two right-invariant Hamiltonians are functionally independent from I_1, I_2, I_3.

It follows that a left-invariant Hamiltonian H is completely integrable on a six dimensional Lie group G_ϵ whenever there is one more integral of motion independent of I_1, I_2, I_3. The second part of this paper provides a complete classification of the integrable cases for such six dimensional Lie groups. For now let us simply note that the extremal equations of the generalized elastic problem on three dimensional space forms can be also expressed in terms of the vector product in $\mathbb{R}^3$ as follows:

$$(41) \qquad \frac{d\hat{M}}{dt} = \hat{M} \times \hat{\Omega} + \hat{P} \times \hat{B}_0 \ , \quad \frac{d\hat{P}}{dt} = \hat{P} \times \hat{\Omega} + \epsilon(\hat{M} \times \hat{B}_0)$$

where

$$\hat{\Omega} = \begin{pmatrix} \frac{1}{\lambda_1} M_1 \\ \frac{1}{\lambda_2} M_2 \\ \frac{1}{\lambda_3} M_3 \end{pmatrix} \ , \quad \hat{M} = \begin{pmatrix} M_1 \\ M_2 \\ M_3 \end{pmatrix} \ , \quad \hat{P} = \begin{pmatrix} p_1 \\ p_2 \\ p_3 \end{pmatrix} \ , \quad B_0 = \begin{pmatrix} b_1 \\ b_2 \\ b_3 \end{pmatrix} \ .$$

The Level set $I_2 = 0$ and the theorem of Poinsot.

Consider now the level set $I_2 = 0$ for the equations (41). In the spherical case $\epsilon = 1$ and $I_2 = \|P(t)\|^2 + \|M(t)\|^2$. Therefore both $P(t)$ and $M(t)$ must be equal to zero. The corresponding elastic curves are the spherical geodesics.

In the Euclidean case the situation is more varied as will be presently demonstrated. First note that

$$(42) \qquad R(t)\hat{P}(t) = \text{constant and} \quad R(t)M(t)R^{-1}(t) + x(t) \wedge R(t)\hat{P}(t) = \text{constant}$$

for the flow of any left-invariant Hamiltonian on the semi direct product $\mathbb{R}^3 \rtimes so_3(\mathbb{R})$. This conservation law may be easily verified through the equations (34).

On the level set $I_2 = 0$, $\hat{P}(t) = 0$, and hence $R(t)M(t)R^{-1}(t) = $ constant. Equivalently, $R(t)\hat{M}(t)$ is constant. This constant, denoted by m is called the angular momentum.

Since $\hat{P}(t) = 0$ the extremal equations reduce to
$$\frac{d\hat{M}}{dt} = \hat{M} \times \hat{\Omega}, \text{ and } \frac{dg(t)}{dt} = g(t)(A_0 + \Omega(t))$$

The rotational part $R(t)$ of the curve $g(t)$ is a solution of $\dfrac{dR(t)}{dt} = R(t)\Omega(t)$. The equations $\dfrac{dR(t)}{dt} = R(t)\Omega(t)$, $\dfrac{d\hat{M}}{dt} = \hat{M} \times \hat{\Omega}(t)$, $J\Omega = M$ coincide with the equations of the top whose center of gravity is at the fixed point of the body. (Top of Euler) Here J denotes a diagonal matrix with its diagonal entries equal to the principal moments of inertia.

The conservation law $R(t)M(t)R^{-1}(t)$ constant forms a basis for a theorem of Poinsot in mechanics. To explain, let

$$E = \left\{ u \in \mathbb{R}^3 : \frac{1}{\lambda_1} u_1^2 + \frac{1}{\lambda_2} u_2^2 + \frac{1}{\lambda_3} u_3^2 = 1 \right\}.$$

The ellipsoid E can be described also in terms of the antisymmetric matrices U as $\{U : \langle JU, U \rangle = 1\}$ where $\langle A, B \rangle = -\frac{1}{2} Trace(AB)$.

Then,

THEOREM 2. *(Poinsot). Suppose that $R(t)$, $x(t)$, $M(t)$ is any extremal curve that satisfies $p_1 = p_2 = p_3 = 0$. The ellipsoid $R(t)E = \{R(t)u : u \in E\}$ rolls without slipping on a plane perpendicular to $m = R(t)\hat{M}(t)$.*

PROOF. Let π_t denote a plane which is tangent to the ellipsoid $E_t = R(t)E$ and which is perpendicular to m. There are exactly two such planes and at the point of tangency the normal vector to E_t is parallel to m. These points of tangency are equal to $\pm\dfrac{R(t)\hat{\Omega}(t)}{\sqrt{H}}$ with $\Omega(t) = J^{-1}M(t)$ for the following reasons. To begin with, $\dfrac{1}{2}(J\Omega(t),\Omega(t)) = H$ and therefore $\dfrac{\hat{\Omega}(t)}{\sqrt{H}}$ belongs to E. Hence $\pm\dfrac{R(t)\hat{\Omega}(t)}{\sqrt{H}}$ belongs to E_t.

Each point $w = R(t)u$, for $u \in E$ satisfies

$$1 = \langle JR^{-1}(t)w, R^{-1}(t)w \rangle = \langle R(t)JR^{-1}(t)w, w \rangle.$$

Hence $E_t = \{w : \langle R(t)JR^{-1}(t)w, w \rangle = 1\}$, and therefore the normal at each point w of E_t is in the direction of $RJR^{-1}w$. In particular the direction of the normal at $w_t = \pm\dfrac{R(t)\hat{\Omega}(t)}{\sqrt{H}}$ is given by

$$\pm\frac{RJR^{-1}R\hat{\Omega}(t)}{\sqrt{H}} = \pm\frac{R(t)J\hat{\Omega}}{\sqrt{H}} = \pm\frac{R(t)\hat{M}(t)}{\sqrt{H}} = \pm\frac{1}{\sqrt{H}}m \,,$$

which is parallel to m, and thus $w_t = \pm\dfrac{R(t)\hat{\Omega}(t)}{\sqrt{H}}$ are the points of tangency with π_t.

The Euclidean inner product $\langle w_t, m \rangle$ is constant for the following reason

$$\langle w_t, m \rangle = \pm\langle\frac{R(t)\hat{\Omega}(t)}{\sqrt{H}}, R(t)\hat{M}(t)\rangle = \pm\frac{1}{\sqrt{H}}\langle\hat{\Omega}(t), \hat{M}(t)\rangle$$
$$= \pm\frac{1}{\sqrt{H}}\langle\hat{\Omega}, J\hat{\Omega}\rangle = \pm\frac{1}{\sqrt{H}}2H = 2\sqrt{H} = \text{constant}.$$

Since $\langle w_t, m \rangle\frac{w_t}{\|w_t\|}$ is the projection of w_t on the line through m it follows that π_t is situated a fixed distance from the origin; that is, it does not vary with time.

Finally $\dfrac{dR}{dt}\hat{\Omega}(t)$ is equal to the speed of the point of contact of E_t with π, and since $\dfrac{dR}{dt}\hat{\Omega} = 0$, E_t rolls without slipping on the plane π.
$\square$

In contrast to the top, however, the elastic problem requires one more integration to get to the complete solution. Namely,

$$x(t) - x(0) = \int_0^t R(\tau) B_0 d\tau$$

is the corresponding elastic curve.

The reader may have already noticed that the conservation law $R(t)M(t)R^{-1}(t) = \text{constant}$ implies $\|\hat{M}(t)\| = \text{constant}$. Hence, the extremal variables $M(t)$ are confined to the intersection of the energy ellipsoid $\mathcal{H} = \frac{1}{2}\langle M, J^{-1}M \rangle$ with the momentum sphere $\langle M, M \rangle = \text{const.}$ As in the case of the top these equations can be integrated explicitly in terms of the suitable coordinates that will be detailed later on in the paper.

In the hyperbolic case $I_2 = 0$ implies that the solutions reside on the light-cone $\|p\|^2 = \|M\|^2$. This case is the most interesting since it is not clear that the corresponding solutions can be integrated by quadrature as in the Euclidean case.

Part II. Complex Lie Groups and complex Hamiltonians

The ultimate understanding of integrable systems associated with the elastic problems and the mechanical tops requires extensions to complex Hamiltonian systems. The remaining part of the paper, a follow up of the ideas initiated by S. Kowalewski in her seminal paper from 1889 ([**Kw**]), demonstrates that $SO_{n+1}(\mathbb{C})$ is the unifying geometric setting for the integrability theory of elastic problems.

VI. Complexified elastic problems

Recall that $G = SO_n(\mathbb{C})$ is the matrix group that leaves the quadratic form $z \cdot w = z_1 w_1 + \cdots + z_n w_n$ in $\mathbb{C}^n$ invariant. Here it is understood that G acts by the matrix multiplication on the elements $z = \begin{pmatrix} z_1 \\ \vdots \\ z_n \end{pmatrix}$ in $\mathbb{C}^n$. It follows that R^{-1} is equal to the matrix transpose of R for each R in $SO_n(\mathbb{C})$. Therefore, the Lie algebra $\mathfrak{g} = so_n(\mathbb{C})$ of G consists of antisymmetric $n \times n$ matrices with complex entries.

To make connections with the elastic problems it will be necessary to induce a Cartan decomposition of $\mathfrak{g}$ which will be done through an involutive automorphism σ defined as follows.

Let D denote the matrix in $SO_{n+1}(\mathbb{C})$ which is diagonal with its diagonal entries $-1, \underbrace{1, \ldots, 1}_{n}$, and then consider the mapping $\sigma : G \to G$ defined by $\sigma(g) = DgD^{-1}$. It is easy to verify that σ is an involutive $(\sigma^2 = I)$ automorphism of G. Its tangent map σ_* at the group identity induces a decomposition of $\mathfrak{g} = \mathfrak{p} + \mathfrak{k}$ with

$$\mathfrak{p} = \{A : \sigma_*(A) = -A\}, \text{ and } \mathfrak{k} = \{A : \sigma_*(A) = A\} \ .$$

Since σ_* is an isomorphism of $\mathfrak{g}$, it follows that $[\mathfrak{p}, \mathfrak{p}] = \mathfrak{k}$, $[\mathfrak{p}, \mathfrak{k}] \subseteq \mathfrak{p}$, and $[\mathfrak{k}, \mathfrak{k}] = \mathfrak{k}$. It follows that $\mathfrak{k}$ is the Lie algebra of the group K consisting of fixed points of σ. Evidently, $K = \{1\} \times SO_n(\mathbb{C})$, and $\mathfrak{k} = \{0\} \times so_n(\mathbb{C})$.

Matrices B in $\mathfrak{p}$ and $\tilde{A}$ in $\mathfrak{k}$ are of the form

(43)
$$B = \begin{pmatrix} 0 & -b^t \\ b & 0 \end{pmatrix} \text{ and } \tilde{A} = \begin{pmatrix} 0 & 0_n^t \\ 0_n & A \end{pmatrix}$$

where $b = \begin{pmatrix} b_1 \\ \vdots \\ b_n \end{pmatrix}$ with $b^t = (b_1 \ldots b_n)$, 0_n is the column consisting of n zeros, and A an antisymmetric $n \times n$ matrix with complex entries.

The fact that $[\mathfrak{p}, \mathfrak{k}] = \mathfrak{p}$ implies that K acts on the vector space $\mathfrak{p}$ by the adjoint action. Hence the vector space of $(n + 1) \times (n + 1)$ antisymmetric matrices can be regarded both as the Lie algebra of SO_{n+1} and the Lie algebra of the semi direct product $G_s = \mathfrak{p} \rtimes K$ through the following arguments.

The group multiplication in G_s is given by $(A, g)(B, h) = (A + gBg^{-1}, gh)$. Hence, $e = (0, I)$ is the group identity, and $(A, g)^{-1} = (-g^{-1}Ag, g^{-1})$ is the group inverse. Then the Lie algebra $\mathfrak{g}_s$ of G_s is the product $\mathfrak{p} \rtimes \mathfrak{k}$ with the Lie bracket structure

$$[(B_1, A_1), (B_2, A_2)]_s = (\mathrm{ad}B_1(A_2) - \mathrm{ad}B_2(A_1), [A_1, A_2])$$

for any (B_1, A_1) and (B_2, A_2) in $\mathfrak{p} \times \mathfrak{k}$.

After $\mathfrak{p}$ is identified with $\mathfrak{p} \times \{0\}$ and $\mathfrak{k}$ with $\{0\} \times \mathfrak{k}$, elements (B, A) in $\mathfrak{p} \times \mathfrak{k}$ are identified with sums $B + A$ in $\mathfrak{g}$ and the Lie brackets can be written as

$$[(B_1, A_1), (B_2, A_2)]_s = ([B_1, A_2] - [B_2, A_1]) + [A_1, A_2] \ .$$

The complex vector space of antisymmetric $(n+1) \times (n+1)$ matrices with complex entries, together with an involution σ_* will be considered either as the Lie algebra $\mathfrak{g} = so_{n+1}(\mathbb{C})$ or as the semi direct product Lie algebra $\mathfrak{g}_s = \mathfrak{p} \rtimes \mathfrak{k}$.

1. The structure of $SO_4(\mathbb{C})$ and the real forms of its Lie algebra.

The Lie algebra $so_4(\mathbb{C})$ together with its real forms has a particularly rich structure that is of considerable importance for the remaining part of the paper. Partly because of the notational necessity and partly as a convenience to the reader the relevant facts will be assembled in the text below. The following proposition is basic.

PROPOSITION 6.1. *(a) $SL_2(\mathbb{C}) \times SL_2(\mathbb{C})$ is a double cover of $SO_4(\mathbb{C})$.*
(b) $sl_2(\mathbb{C}) \times sl_2(\mathbb{C})$ is isomorphic to $so_4(\mathbb{C})$.

PROOF. To each z in $\mathbb{C}^4$ associate the matrices

$$Z = \begin{pmatrix} z_0 + iz_1 & z_2 + iz_3 \\ -z_2 + iz_3 & z_0 - iz_1 \end{pmatrix} \text{ and } Z^\dagger = \begin{pmatrix} z_0 - iz_1 & -z_2 - iz_3 \\ z_2 - iz_3 & z_0 + iz_1 \end{pmatrix}.$$

The reader can easily verify that the operation $Z \to Z^\dagger$ is an anti-automorphism and that $Z^\dagger = Z^{-1}$ for $Z \in SL_2(\mathbb{C})$. Moreover,

$$(44) \qquad \frac{1}{2}(ZW^\dagger + WZ^\dagger) = (z \cdot w)I$$

Let $\Phi : SL_2(\mathbb{C}) \times SL_2(\mathbb{C}) \to M_4(\mathbb{C})$ be defined as follows

$$(45) \qquad \Phi(g_1, g_2)z = w \text{ if and only if } g_1 Z g_2^{-1} = W$$

for all $z \in \mathbb{C}^4$. It follows that Φ is a homomorphism whose kernel is $\pm(I, I)$. In addition $\Phi(g_1, g_2) = R$ satisfies $Rz \cdot Rw = z \cdot w$ as can be easily verified through equation (44), and therefore the range of Φ is equal to $SO_4(\mathbb{C})$.

The tangent map Φ_* at the identity provides the isomorphism between the Lie algebras. It follows that $(U, V) \in sl_2(\mathbb{C}) \times sl_2(\mathbb{C})$ corresponds to $C \in so_4(\mathbb{C})$ whenever

$$(46) \qquad UZ - ZV = W \text{ if and only if } Cz = w$$

for all $z \in \mathbb{C}^4$. To get explicit formulas out of (46) it is useful and convenient to introduce the following matrices:

$$(47) \qquad E_1 = \begin{pmatrix} i & 0 \\ 0 & -i \end{pmatrix}, E_2 = \begin{pmatrix} 0 & 1 \\ -1 & 0 \end{pmatrix}, E_3 = \begin{pmatrix} 0 & i \\ i & 0 \end{pmatrix}.$$

The above matrices are isomorphic to the standard quaternions i, j, k in the sense that they conform to the relations:

$$E_1^2 = E_2^2 = E_3^2 = -I, \text{ and } E_1 E_2 = E_3, E_3 E_1 = E_2, E_2 E_3 = E_1.$$

Then $Z = z_0 I + z_1 E_1 + z_2 E_2 + z_3 E_3$ corresponds to a point z in $\mathbb{C}^4$ with coordinates z_0, z_1, z_2, z_3. It will be also convenient to maintain the conventions established earlier and use $\hat{B}$ and $\hat{A}$ to denote the vectors in $\mathbb{C}^3$ that correspond to matrices B in $\mathfrak{p}$ and A in $\mathfrak{k}$. With all these notations at our disposal, let $C = A + B$ with $A \in \mathfrak{k}$ and $B \in \mathfrak{p}$, and let

$$\hat{A} = \begin{pmatrix} a_1 \\ a_2 \\ a_3 \end{pmatrix}, \hat{B} = \begin{pmatrix} b_1 \\ b_2 \\ b_3 \end{pmatrix}.$$

In addition, let $U = \sum u_i E_i$ and let $V = \sum v_i E_i$. Then relations (46) imply that

$$u_1 = \frac{1}{2}(a_1 + b_1), u_2 = \frac{1}{2}(a_2 + b_2), u_3 = \frac{1}{2}(a_3 + b_3)$$

(48)
$$v_1 = \frac{1}{2}(a_1 - b_1), v_2 = \frac{1}{2}(a_2 - b_2), v_3 = \frac{1}{2}(a_3 - b_3).$$

Therefore,

(49) $\quad U + V = \begin{pmatrix} ia_1 & a_2 + ia_3 \\ -a_2 + ia_3 & -ia_1 \end{pmatrix}, U - V = \begin{pmatrix} ib_1 & b_2 - ib_3 \\ -b_2 - ib_3 & -ib_1 \end{pmatrix}$

COROLLARY. $SL_2(\mathbb{C})$ *is a double cover of* $SO_3(\mathbb{C})$. *Consequently,* $sl_2(\mathbb{C}) = so_3(\mathbb{C})$.

PROOF. $SO_3(\mathbb{C}$ is isomorphic to the isotropy subgroup $Ke_0 = e_0$ of $SO_4(\mathbb{C})$. Let Φ denote the homomorphism defined and used in the previous proposition. It is easy to verify that $\Phi(g_1, g_2)$ is in K if and only if $g_1 = g_2$. Therefore $SL_2(\mathbb{C})$ is a double cover of K and hence of $SO_3(\mathbb{C})$.

Then it follows from (48) and (49) that a matrix $A \in so_3(\mathbb{C})$ corresponds to the matrix $U \in sl_2(\mathbb{C})$ if and only if

$$U = \frac{1}{2} \begin{pmatrix} ia_1 & (a_2 + ia_3) \\ -(a_2 + ia_3) & -ia_1 \end{pmatrix}.$$

DEFINITION 6.1. *A real Lie algebra $\mathfrak{g}_0$ is said to be a real form for a complex Lie algebra $\mathfrak{g}$ if*

$$\mathfrak{g} = \mathfrak{g}_0 + i\mathfrak{g}_0 .$$

Evidently, $so_n(\mathbb{R})$ is a real form for the complex Lie algebra $so_n(\mathbb{C})$. It is also well known that each real algebra $so(p,q)$ with $n = p + q$ is a real form for $so_n(\mathbb{C})$. In particular, $so(1,2)$, $so(2,1)$, and $so_3(\mathbb{R})$ are all real forms for $so_3(\mathbb{C})$. To show the connections between complex Lie groups and the elastic problems it will be necessary to discuss the real forms in more detail.

Let V_{-1} denote the real vector space spanned by ie_0, e_1, e_2, e_3. Then $V_{-1} \subset \mathbb{C}^4$, and $x \cdot y = -x_0 y_0 + \sum x_i y_i$ for any $x = ix_0 e_0 + \sum x_i e_i$ and $y = iy_0 e_0 + \sum y_i e_i$ in V_{-1}. It is easy to verify that V_{-1} is an invariant subspace for a matrix $L = iB + A$ with $B \in \mathfrak{p}$ and $A \in \mathfrak{k}$. Hence the restriction of L to V_{-1} is an element of $so(1,3)$.

Let now A_1, A_2, A_3 and B_1, B_2, B_3 denote any bases in $\mathfrak{k}$ and $\mathfrak{p}$ that are orthonormal relative to the trace form $\langle L, M \rangle = -\frac{1}{2}\text{Trace}(LM)$ and satisfy the following Lie bracket table:

$[\,,\,]$	A_1	A_2	A_3	B_1	B_2	B_3
A_1	0	$-A_3$	A_2	0	$-B_3$	B_2
A_2	A_3	0	$-A_1$	B_3	0	$-B_1$
A_3	$-A_2$	A_1	0	$-B_2$	B_1	0
B_1	0	$-B_3$	B_2	0	$-A_3$	A_2
B_2	B_3	0	$-B_1$	A_3	0	$-A_1$
B_3	$-B_2$	B_1	0	$-A_2$	A_1	0

TABLE 1

Such bases can be constructed as follows. Recall the following property of the trace form

$$\langle L, M \rangle = \hat{A}_1 \cdot \hat{A}_2 + \hat{B}_1 \cdot \hat{B}_2$$

where $L = A_1 + B_1$ and $M = A_2 + B_2$ are the decompositions into $\mathfrak{k}$ and $\mathfrak{p}$ factors, and where $z \cdot w = z_1 w_1 + z_2 w_2 + z_3 w_3$ denotes the complex inner product in $\mathbb{C}^3$.

Vectors a, b, c in $\mathbb{C}^3$ are said to form a right-handed triad if $a \cdot a = b \cdot b = c \cdot c = 1$ and $a \times b = c$, $c \times a = b$ and $b \times c = a$ with $\times$ equal to the vector product in $\mathbb{C}^3$.

Any orthonormal basis A_1, A_2, A_3 in $\mathfrak{k}$ (relative to the trace norm) can be renumbered so that $\hat{A}_1$, $\hat{A}_2$, $\hat{A}_3$ form a right-handed orthonormal triad. In such a case A_1, A_2, A_3 satisfy the following Lie bracket table:

$$[A_1, A_2] = -A_3 \ , \ [A_1, A_3] = A_2 \ \text{ and } [A_2, A_3] = -A_1 \ .$$

Recall that $\widehat{[A_i, A_j]} = \hat{A}_j \times \hat{A}_i$, from which the above relations follow. An orthonormal basis A_1, A_2, A_3 such that $\hat{A}_1$, $\hat{A}_2$, $\hat{A}_3$ is a right-handed triad will be extended to a basis A_1, A_2, A_3, B_1, B_2, B_3 in $\mathfrak{g}$ such that $\hat{B}_i = \hat{A}_i$, $i = 1, 2, 3$. Then, $[B_1, B_2] = -A_3$, $[B_1, B_3] = A_2$ and $[B_2, B_3] = -A_1$, and similarly $[A_1, B_2] = -B_3$, $[A_1, B_3] = B_2$ and $[A_2, B_3] = -B_1$, $[A_2, B_1] = B_3$, $[A_3, B_1] = -B_2$ $[A_3, B_2] = B_1$.

The last relations follow from the fact that $\widehat{[A_i, B_j]} = -A_i \hat{B}_j = -\hat{A}_i \times \hat{B}_j$, and that $\hat{B}_j = \hat{A}_j$.

Considered as a basis for the semi-direct Lie algebra $\mathfrak{g}_s$, the preceding basis conforms to the following Lie bracket table:

$[\ ,\ ]_s$	A_1	A_2	A_3	B_1	B_2	B_3
A_1	0	A_3	A_2	0	$-B_3$	B_2
A_2	$-A_3$	0	$-A_1$	B_3	0	$-B_1$
A_3	$-A_2$	A_1	0	$-B_2$	B_1	0
B_1	0	$-B_3$	B_2	0	0	0
B_2	B_3	0	$-B_1$	0	0	0
B_3	$-B_2$	B_1	0	0	0	0

TABLE 2

Then the restriction to V_{-1} of the real vector space spanned by A_1, A_2, A_3, iB_1, iB_2, iB_3 is equal to $so(1,3)$, and therefore $so(1,3)$ is a real form for $so_4(\mathbb{C})$. Of course, $so_4(\mathbb{R})$ is the real Lie algebra generated by A_1, A_2, A_3, B_1, B_2, B_3.

We can parameterize all these real Lie algebras by a single parameter ε with $\epsilon = 0$ corresponding to the semi-direct product, $\epsilon = -1$ corresponding to $so(1,3)$ and $\epsilon = 1$ corresponding to $so_4(\mathbb{R})$. Then the following table covers all three cases

$[\,,\,]_\epsilon$	A_1	A_2	A_3	$B_1(\epsilon)$	$B_2(\epsilon)$	$B_3(\epsilon)$
A_1	0	$-A_3$	A_2	0	$B_3(\epsilon)$	$B_3(\epsilon)$
A_2	A_3	0	$-A_1$	$B_3(\epsilon)$	0	$B_2(\epsilon)$
A_3	$-A_2$	A_1	0	$-B_2(\epsilon)$	$-B_1(\epsilon)$	0
$B_1(\epsilon)$	0	$-B_3(\epsilon)$	$B_2(\epsilon)$	0	$-\epsilon A_3$	ϵA_2
$B_2(\epsilon)$	$B_3(\epsilon)$	0	$-B_1(\epsilon)$	ϵA_3	0	$-\epsilon A_1$
$B_3(\epsilon)$	$-B_2(\epsilon)$	$B_1(\epsilon)$	0	$-\epsilon A_2$	ϵA_1	0

TABLE 3

where $B_j(\epsilon) = \begin{cases} B_j & \text{for } \epsilon = 0, \epsilon = 1 , \\ iB_j & \text{for } \epsilon = -1 \end{cases}$ $j = 1, 2, 3.$

However, $so_4(\mathbb{C})$ has another distinctive real form, namely $so(2,2)$. This real Lie algebra is generated by A_1, iA_2, iA_3, B_1, iB_2, iB_3. The corresponding Lie bracket table is given by Table 4 below.

	A_1	iA_2	iA_3	B_1	iB_2	iB_3
A_1	0	$-iA_3$	iA_2	0	iB_3	iB_2
iA_2	iA_3	0	A_1	iB_3	0	B_1
iA_3	$-iA_2$	$-A_1$	0	$-iB_2$	$-B_1$	0
B_1	0	$-iB_3$	iB_2	0	$-iA_3$	iA_2
iB_2	iB_3	0	B_1	iA_3	0	A_1
iB_3	$-iB_2$	$-B_1$	0	$-iA_2$	$-A_1$	0

TABLE 4

The semi-direct algebra $\mathfrak{g}_s$ has one more real form namely $\mathbb{R}^3 \ltimes so(2,1)$. It is the real Lie algebra generated by the preceding basis relative to the semi-direct Lie bracket.

PROPOSITION 6.2. $SL_2(\mathbb{C})$ *is a double cover of* $SO(1,3)$, *and* $SU(1,1) \times SU(1,1)$ *is a double cover of* $SO(2,2)$.

PROOF. Let V_{-1} denote the real linear span of ie_0, e_1, e_2, e_3. For each $z = ix_0 e_0 + x_1 e_1 + x_2 e_2 + x_3 e_3$ in V_{-1} the matrix Z is of the form

$$Z = i \begin{pmatrix} x_0 + x_1 & x_3 - ix_2 \\ x_3 + ix_2 & x_0 - x_1 \end{pmatrix}.$$

Then $g_1 Z g_2^{-1} = W$ with z and w in V_{-1} if and only if $g_2^{-1} = g_1^*$. The verification of this detail will be left to the reader.

Then $\Phi : SL_2(\mathbb{C}) \to SO(1,3)$ defined by $\Phi(g) = R$ through the relation

$$gZg^* = W \text{ whenever } Rz = w$$

for $z \in V_{-1}$ is the desired homomorphism.

The proof for $SO(2,2)$ is similar and will be omitted.

It is a corollary that $sl_2(\mathbb{C})$ and $so(1,3)$ are isomorphic. The isomorphism is given by the tangent map Φ_* of Φ at the identity. It follows that $C \in so(1,3)$ corresponds to $U \in sl_2(\mathbb{C})$ if and only if

$$UZ - ZU^* = W \text{ whenever } Cz = w.$$

In particular, matrices $B \in \mathfrak{p}$ correspond to the Hermitian matrices in $sl_2(\mathbb{C})$, while matrices A in $\mathfrak{k}$ correspond to the skew-Hermitian matrices in $sl_2(\mathbb{C})$.

VII. Complex elasticae of Euler and its n-dimensional extensions

Let σ denote the involutive automorphism of $G = SO_{n+1}(\mathbb{C})$ defined in the previous section such that the corresponding decomposition $\mathfrak{g} = so_{n+1}(\mathbb{C}) = \mathfrak{p} + \mathfrak{k}$ is described according to the equation (43). Then $\mathfrak{k}$ is isomorphic to $so_n(\mathbb{C})$, and matrices

$$A_{ij} = e_i \wedge e_j,\ 1 \le i \le n,\ 1 \le j \le n$$

form an orthonormal basis for $\mathfrak{k}$ relative to the trace form. Matrices $E_i = \begin{pmatrix} 0 & -e_i^t \\ e_i & 0 \end{pmatrix}$, $i = 1,2\dots,n$ form an orthonormal basis for $\mathfrak{p}$, and together with the above matrices form an orthonormal basis in $\mathfrak{g}$.

Then any ℓ in $\mathfrak{g}^*$ can be expressed as

$$\ell = \sum p_i E_i^* + \sum M_{ij} A_{ij}^*$$

relative to the dual basis $\{E_i^*,\ i = 1, \dots, n\}$ and $\{A_{ij}^*,\ i = 1, \dots, n-1, j = 1, \dots, n-1\}$.

We will now consider the Hamiltonian equations defined by the left-invariant complex Hamiltonian H on $\mathfrak{g}^*$ given by

(50)
$$\mathcal{H} = \frac{1}{2}\sum_{i=1}^{n-1} M_{1i}^2 + p_1$$

The above Hamiltonian is the complex version of the Hamiltonian H on G_ϵ corresponding to the Euler-Griffiths problems defined by (equation (11)).

To be consistent with the notations used earlier let $A_1 = e_1 \wedge e_2, A_2 = e_1 \wedge e_3, \dots, A_{n-1} = e_1 \wedge e_n$, and let $M_1 = M_{12}, \dots, M_{n-1} =$

M_{1n}. The Hamiltonian equations of $\mathcal{H}$ can be written on $\mathfrak{g}$ according to (32) as

(E_1) $\dfrac{dM}{dt}(t) = [\Omega(t), M(t)] + [E_1, P(t)], \dfrac{dP}{dt}(t) = [\Omega(t), P(t)] + [E_1, M(t)],$

and $\frac{dg}{dt}(t) = g(t)(E_1 + \Omega(t))$ where $\Omega(t) = \sum M_i(t)A_i$.

But $\mathcal{H}$ can be also considered as a left-invariant Hamiltonian on the semi direct product $\mathfrak{p} \rtimes SO_n(\mathbb{C})$ in which case its Hamiltonian equations have the same form as the equations (35):

(E_0)
$\dfrac{dM}{dt}(t) = [\Omega(t), M(t)] + e_1 \wedge \hat{P}(t), \dfrac{d\hat{P}}{dt}(t) = \Omega(t)\hat{P}(t)$ and $\dfrac{dg}{dt}(t) = g(t)(E_1 + \Omega(t))$

where $\Omega(t) = \sum M_i(t)A_i$.

The 3-dimensional case.

Since $K = SO_2(\mathbb{C})$ is a commutative group, $[\Omega, M] = 0$. Then equations (E_0) and (E_1) can be written explicitly in terms of a parameter ϵ as:

(Elast) $\dfrac{dM_1}{dt} = p_2 \, , \dfrac{dp_1}{dt} = M_1 p_2 \, , \dfrac{dp_2}{dt} = M_1 p_1 - \epsilon M_1$

with ϵ indicating the Lie algebra $\mathfrak{g}_\epsilon$.

As remarked in Section V, all three dimensional left-invariant Hamiltonian systems are completely integrable. In this case every solution curve of (Elast) is contained in the intersection of two complex surfaces $I_1 = \mathcal{H} = \dfrac{1}{2}M_1^2 + p_1$ and $I_2 = p_1^2 + p_2^2 + \epsilon M_1^2$. for suitable complex numbers I_1 and I_2.

Let $S(I_1, I_2)$ denote this intersection. Points $z = (p_1, p_2, M_1)$ of S are called *smooth* if the differentials dI_1 and dI_2 are linearly independent at z. Otherwise they are called *singular*. It follows that the singular points of S are the points at which vectors $(1, 0, M_1)$ and $(p_1, p_2, \epsilon M_1)$ are linearly dependent i.e., points z of $S(I_1, I_2)$ defined by $M_1(\epsilon - p_1) = 0$ and $p_2 = 0$. It follows that the singular points coincide with the equilibrium points of (Elast).

At each smooth point $S(I_1, I_2)$ is locally a complex manifold. We shall presently show that $S(I_1, I_2)$ is a double cover of an elliptic curve at each smooth point.

Consider the following transformation from $S(I_1, I_2)$ into $\mathbb{C}^2$: $\eta = \dfrac{1}{2}M_1 p_2$ and $u = -\dfrac{1}{4}M_1^2$. At smooth points $M_1 \neq 0$, and therefore the preceding transformation is holomorphic. Moreover, every point (u, η)

with $u \neq 0$ corresponds to exactly two points (p_1, p_2, M_1) on $S(I_1, I_2)$ given by

$$p_1 = I_1 + 2u \quad , \quad p_2 = \frac{2\eta}{\sqrt{-4u}} \quad , \quad M_1 = \sqrt{-4u}$$

$$p_1 = I + 2u \quad , \quad p_2 = -\frac{2\eta}{\sqrt{-4u}} \quad , \quad M_1 = -\sqrt{-4u}$$

Furthermore,

$$\eta^2 = \frac{1}{4} M_1^2 p_2^2 = -u(I_2 - p_1^2 - \epsilon M_1^2) = -u\left(I_2 - (I_1 - \frac{1}{2} M_1^2)^2 - \epsilon M_1^2\right)$$

$$= -u\left(-\frac{1}{4} M_1^4 + M_1^2(I_1 - \epsilon) + (I_2 - I_1^2)\right)$$

Hence,

$$\eta^2 = 4u^3 + 4u^2(I_1 - \epsilon) - u(I_2 - I_1^2) .$$

The variable $\xi = u - \frac{1}{3}(I_1 + \epsilon)$ reduces the preceding elliptic curve to the elliptic curve of Weierstrass

(W) $$\eta^2 = 4\xi^3 - g_2\xi - g_3$$

with

$$g_2 = \frac{4}{3}(I_1 - \epsilon)^2 + (I_2 - I_1)^2 \text{ and } g_3 = -\frac{8}{27}(I_1 - \epsilon)^3 - \frac{1}{3}(I_1 - \epsilon)(I_2 - I_1^2) .$$

It is easy to show that any complex parameters g_2 and g_3 can be realized by suitable complex numbers I_1 and I_2.

Let now $p_1(t)$, $p_2(t)$, $M_1(t)$ denote any integral curve of (Elast), and let

$$\xi(t) = -\frac{1}{4} M_1^2(t) - \frac{1}{3}(I_1 - \epsilon).$$

Then $\frac{du}{dt} = \frac{d}{dt}\left(-\frac{1}{4} M_1^2(t)\right) = -\frac{1}{2} M_1(t) \frac{dM_1}{dt} = -\frac{1}{2} M_1(t) p_2(t)$, and therefore

$$\left(\frac{du}{dt}\right)^2 = \frac{1}{4} M_1^2(t) p_2^2(t) = \eta^2(t) = 4\xi^3(t) - g_2\xi(t) - g_3$$

Since $\frac{du}{dt} = \frac{d\xi}{dt}$ we have shown that

(51) $$\left(\frac{d\xi}{dt}\right)^2 = 4\xi^3(t) - g_2\xi(t) - g_3 \text{ for all } t .$$

The solutions of equation (51) are well known in the literature on elliptic functions ([**Sg**]). Every solution curve of (51) is a meromorphic function of complex time having a single pole at a point t_0 determined by the initial value $\xi(0) = \xi_0$. It is convenient to shift the pole to $t_0 = 0$, in which case $\xi(t)$ is equal to the $\wp$-function of Weierstrass. This function is an even doubly periodic meromorphic function having a double pole at $z = 0$. The periods of $\wp$ are determined by the constants g_2 and g_3.

Thus every non-stationary solution $p_1(t), p_2(t), M(t)$ of (Elast) is parametrized by $\wp$ with

$$-\frac{1}{4}M_1^2(t) = \wp(t - t_0) + \frac{1}{3}(I_1 - \epsilon)$$

$$p_1(t) = I_1 - \frac{1}{2}M_1^2(t) \quad \text{and}$$

$$p_2(t) = \sqrt{I_2 - p_1^2(t) - \epsilon M_1^2(t)} \ .$$

It follows that each solution $P(t), M(t)$ of (Elast) is a meromorphic function of complex time. To integrate the extremal equations on the entire cotangent bundle of the underlying group G, a choice has to be made. Any function on $\mathfrak{g}$ may be regarded as a left-invariant Hamiltonian on any one of the groups $G = \mathfrak{p} \rtimes SO_2(\mathbb{C})$, $G = SO_3(\mathbb{C})$ or $G = SL_2(\mathbb{C})$. We will omit the details concerning the solutions in the semi-direct case and $SO_3(\mathbb{C})$. They are formally the same as in [**Ju1**]. Instead the integration procedure will be carried out in $G = SL_2(\mathbb{C})$.

Each g in $SL_2(\mathbb{C})$ is identified with a point z in $S_c^3 = \{z \in \mathbb{C}^4 : z_0^2 + z_1^2 + z_2^2 + z_3^2 = 1\}$. The identification is provided via the formula

$$g = \begin{pmatrix} z_0 + iz_1 & z_2 + iz_3 \\ -z_2 + iz_2 & z_0 - iz_1 \end{pmatrix}.$$

Then $SL_2(\mathbb{C})$ acts on itself by the adjoint action and the orbit through any $g = \begin{pmatrix} i\hat{z}_1 & \hat{z}_2 + i\hat{z}_3 \\ -\hat{z}_2 + i\hat{z}_2 & -i\hat{z}_1 \end{pmatrix}$ is the complex variety $S_c^2 = \{z \in \mathbb{C}^3 : z_1^2 + z_2^2 + z_3^2 = 1\}$. Therefore S_c^2 can be realized as the quotient $SL_2(\mathbb{C})/K$ with K the subgroup of $SL_2(\mathbb{C})$ that fixes $g_0 = \begin{pmatrix} i & 0 \\ 0 & -i \end{pmatrix}$. It follows that K consists of matrices $g = \begin{pmatrix} a + ib & 0 \\ 0 & a - ib \end{pmatrix}$ that satisfy $a^2 + b^2 = 1$. Evidently K is isomorphic to $SO_2(\mathbb{C})$.

Since S_c^2 contains both the sphere $S^2(\mathbb{R})$ and the hyperboloid $\mathbb{H}^2$ as the real sub varieties, both of which can be obtained as the quotients of real Lie groups that are the exponential groups of real forms for $sl_2(\mathbb{C})$, it seems natural to call the projections of the extremal curves of $\vec{\mathcal{H}}$ onto S_c^2 complex elasticae.

We shall take advantage of the decomposition $SL_2(\mathbb{C}) = KSO_2(\mathbb{C})K$ and parametrize matrices g in $SL_2(\mathbb{C})$ in terms of the complex coordinates $\theta_1, \theta_2, \theta_3$ defined by

$$g = e^{\theta_1 A_1} e^{\theta_2 A_2} e^{\theta_3 A_1}$$

where

$$A_1 = \frac{1}{2} \begin{pmatrix} i & 0 \\ 0 & -i \end{pmatrix}, \; A_2 = \frac{1}{2} \begin{pmatrix} 0 & 1 \\ -1 & 0 \end{pmatrix}, \; A_3 = \frac{1}{2} \begin{pmatrix} 0 & i \\ i & 0 \end{pmatrix}$$

are the Pauli matrices. It follows that

$$e^{\theta_1 A_1} = \begin{pmatrix} e^{\frac{i}{2}\theta_1} & 0 \\ 0 & e^{-\frac{i}{2}\theta_1} \end{pmatrix}, \; e^{\theta_2 A_2} = \begin{pmatrix} \cos\frac{\theta_2}{2} & \sin\frac{\theta_2}{2} \\ -\sin\frac{\theta_2}{2} & \cos\frac{\theta_2}{2} \end{pmatrix}, \; e^{\theta_3 A_1} = \begin{pmatrix} e^{\frac{1}{2}i\theta_3} & 0 \\ 0 & e^{-\frac{1}{2}i\theta_3} \end{pmatrix}.$$

With these background remarks and notations, the integration procedure is carried out as follows. Recall that each extremal curve $\big(g(t), L(t)\big)$ satisfies $g(t)L(t)g^{-1}(t) = \Lambda$ with Λ an element in $sl_2(\mathbb{C})$. Then $g(t)$ may be conjugated by an element in $SL_2(\mathbb{C})$ so that $\Lambda = aA_1$ for some complex number a.

It follows from equations (49) that $L(t) = \frac{1}{2}\begin{pmatrix} M_1(t)i & p_2(t) + ip_1(t) \\ -p_2(t) + ip_1(t) & -M_1(t)i \end{pmatrix}$ and $L(t) = g^{-1}(t)\Lambda g(t)$ reduces to

$$\begin{pmatrix} M_1 i & p_2 + ip_1 \\ -p_2 + ip_1 & -M_1 i \end{pmatrix} = ia \begin{pmatrix} \cos\theta_2 & e^{-i\theta_3}\sin\theta_2 \\ e^{i\theta_3}\sin\theta_2 & -\cos\theta_2 \end{pmatrix}.$$

Therefore, $a^2 = M_1^2(t) + h_1^2(t) + h_2^2(t) = I_2$ and

$$M_1(t) = \sqrt{I_2}\cos\theta_2(t),$$

$$p_2(t) + ip_1(t) = i\sqrt{I_2}\big(\cos\theta_3(t) - i\sin\theta_3(t)\big)\sin\theta_2(t) ,$$

$$-p_2(t) + ip_1(t) = i\sqrt{I_2}\big(\cos\theta_3(t) + i\sin\theta_3(t)\big)\sin\theta_2(t) .$$

Hence,

(52) $\qquad p_2(t) = \sqrt{I_2}\sin\theta_3(t)\sin\theta_2(t)$ and $p_1(t) = \sqrt{I_2}\cos\theta_3(t)\sin\theta_2(t) .$

The above equation provides an explicit relation between the functions $(p_1(t), p_2(t))$ and $(\theta_1(t), \theta_2(t))$. Since both $p_1(t)$ and $p_2(t)$ are meromorphic so are $\theta_2(t)$ and $\theta_3(t)$.

The remaining angle $\theta_1(t)$ will be obtained from the equation

$$\frac{dg}{dt}(t) = g(t)(A_3 + M_1(t)A_1)$$

as follows:

Since $g(t) = e^{A_1\theta_1} e^{A_2\theta_2} e^{A_1\theta_3}$,

$$\frac{dg}{dt}(t) = \dot{\theta}_1 A_1 g(t) + \dot{\theta}_2 e^{A_1\theta_1} A_2 e^{A_2\theta_2} e^{A_1\theta_3} + \dot{\theta}_3 g(t)A_1 ,$$

and hence

(53) $$g^{-1}(t)\frac{dg}{dt}(t) = \dot{\theta}_1 g^{-1}(t)A_1 g(t) + \dot{\theta}_2 e^{-A_1\theta_3}A_2 e^{A_1\theta_3} + \dot{\theta}_3 A_1 \ .$$

Since

$$g^{-1}(t)A_1 g(t) = \frac{1}{\sqrt{I_2}}L(t) = \frac{1}{\sqrt{I_2}}(M_1(t)A_1 + p_2(t)A_2 + p_1(t)A_3)$$

and

$$e^{-A_1\theta_3}A_2 e^{A_1\theta_3} = \cos\theta_3 A_2 - \sin\theta_3 A_3,$$

equation (53) can be rewritten as

$$g^{-1}(t)\frac{dg}{dt} = \frac{\dot{\theta}_1}{\sqrt{I_1}}(M_1 A_1 + p_2 A_2 + p_1 A_3) + \dot{\theta}_2(\cos\theta_3 A_2 - \sin\theta_3 A_3) + \dot{\theta}_3 A_1$$
$$= A_3 + M_1 A_1 \ .$$

It follows that

$$\frac{\dot{\theta}_1}{\sqrt{I_2}}M_1 + \dot{\theta}_3 = M_1 \ , \quad \frac{\dot{\theta}_1}{\sqrt{I_2}}p_2 + \dot{\theta}_2\cos\theta_3 = 0 \ , \quad \text{and} \quad \frac{\dot{\theta}_1}{\sqrt{I_2}}p_1 - \dot{\theta}_2\sin\theta_3 = 1 \ .$$

Then,

$$\frac{1}{\sqrt{I_2}}\dot{\theta}_1 p_2\sin\theta_3 + \dot{\theta}_2\sin\theta_3\cos\theta_3 = 0$$

$$\frac{1}{\sqrt{I_2}}\dot{\theta}_1 p_1\cos\theta_3 - \dot{\theta}_2\sin\theta_3\cos\theta_3 = \cos\theta_3$$

and therefore

$$\frac{1}{\sqrt{I_2}}\dot{\theta}_1(p_2\cos\theta_3 + p_1\sin\theta_3) = \cos\theta_3 \ .$$

So,

$$\dot{\theta}_1 = \frac{\sqrt{I_2}\cos\theta_3}{p_2\cos\theta_3 + p_1\sin\theta_3} \ .$$

But $p_2 = \sqrt{I_2}\sin\theta_3\sin\theta_2$ and $p_1 = \sqrt{I_2}\cos\theta_3\sin\theta_2$, and therefore, $p_2\cos\theta_3 + p_1\sin\theta_3 = \sqrt{I_2}\sin\theta_2$. Hence,

$$\dot{\theta}_1 = \frac{\cos\theta_3}{\sin\theta_2} \ .$$

Therefore,

$$\theta_1(t) = \int_0^t \frac{\cos\theta_3}{\sin\theta_2}dt$$

provided $\dfrac{\cos\theta_3}{\sin\theta_2}$ is residue free.

To show that the above integrand is residue free, first note that

$$\cos\theta_3 = \frac{p_1}{\sqrt{I_2}\sin\theta_2}, \text{ and that } p_1^2 + p_2^2 = I_2\sin^2\theta_2.$$

Therefore,

$$\frac{\cos\theta_3}{\sin\theta_2} = \frac{p_1}{\sqrt{I_2}\sin^2\theta_2} = \frac{\sqrt{I_2}p_1}{p_1^2 + p_2^2} = \frac{\sqrt{I_2}(I_1 - \frac{1}{2}M_1^2)}{I_2 - M_1^2}.$$

Since $M_1^2(t) = -4\wp(t + t_0) + constant$, $M_1^2(t)$ is an even function. Therefore,

$$f(t) = \frac{I_1 - \dfrac{1}{2}M_1^2(t)}{I_2 - M_1^2(t)} \text{ is an even function, hence residue free. Con-}$$

sequently

$$\theta_1(t) = \sqrt{I_2}\int_0^t \frac{(I_1 - \dfrac{1}{2}M_1^2(z))}{I_2 - M_1^2(z)}dz$$

is a meromorphic function of t.

The preceding developments show that each extremal curve $g(t)$ is a meromorphic complex curve. The complex elasticae are the projections of the extremal curves. It follows that they are given by $Y(t) = g(t)\begin{pmatrix} i & 0 \\ 0 & -i \end{pmatrix}g^{-1}(t)$. Since

$$Y(t) = e^{A_1\theta_1(t)}e^{A_2\theta_2(t)}\begin{pmatrix} i & 0 \\ 0 & -i \end{pmatrix}e^{-A_2\theta_2(t)}e^{-A_1\theta_3(t)}$$

$$= \begin{pmatrix} i\cos\theta_2 & (-i\cos\theta_1 + \sin\theta_1)\sin\theta_2 \\ -(i\cos\theta_1 + \sin\theta_1)\sin\theta_2 & -i\cos\theta_2 \end{pmatrix},$$

the complex elasticae are given by the following formulae:

$$y_1(t) = \cos\theta_2(t) \qquad y_3(t) = -\cos\theta_1(t)\sin\theta_2(t) \qquad y_2(t) = \sin\theta_1(t)\sin\theta_2(t).$$

These expressions coincide with the formulas obtained in [Ju1] for the spherical case (except for the renaming of the coordinates $x_3 = y_1$, $x_2 = -y_2$, $x_1 = -y_3$).

The Hamiltonian system considered above is a canonical representative for the Hamiltonians on $sl_2^*(\mathbb{C})$ that are quadratic on su_2^* and linear on $\mathfrak{p}^*$, in the sense that the most general Hamiltonian

$$\mathcal{H} = \frac{1}{2\lambda}M_1^2 + b_1p_1 + b_2p_2$$

with $|b_1|^2 + |b_2|^2 > 0$ and $\lambda > 0$ can be brought to the canonical form $\mathcal{H} = \frac{1}{2}M_1^2 + p_1$ under a symplectic transformation in the cotangent bundle of $SL_2(\mathbb{C})$.

Solutions to the n-dimensional Euler-Griffiths elastic problems. Return now to the general case described by the equations (E_0)

and (E_1). Let $\mathfrak{k}_0$ denote the linear span of $\{A_{ij} = e_i \wedge e_j,\, i \geq 2,\, j \geq 2\}$, and let

$$M = \Omega + \Delta$$

where Ω is as in (E_0) and (E_1), and where Δ is the projection of M on $\mathfrak{k}_0$.

It follows that

$$[E_1, P] = -\sum_{i=2}^{n-1} p_i A_i \text{ and } [E_1, \Delta] = 0,$$

and therefore the extremal equations (E_0) and (E_1) can be written as:

(54)
$$\frac{d\Omega}{dt}(t) = [E_1, P(t)],\ \frac{d\Delta}{dt}(t) = 0,\ \frac{d\hat{P}}{dt}(t) = \Omega(t)(\hat{P}(t) - \epsilon e_1)$$

with $\epsilon = 0$ denoting the semi direct case. It follows that $\Delta(t)$ is constant.

According to the Maximum Principle the extremals to the Euler-Griffiths problem 2 conform to the transversality conditions, but apart from this distinction the Hamiltonian systems are the same for both problems. The transversality conditions mean that Δ should vanish at $t = 0$ and $t = T$. Since Δ is constant it follows that $\Delta = 0$ for problem 2.

The solutions of the above differential system are most naturally expressed in terms of the geometric invariants of the projected curve $z(t) = \pi(g(t)) = g(t)e_0$ on the complex manifold

$$S_c^n = \{z \in \mathbb{C}^{n+1} : z_0^2 + \cdots + z_n^2 = 1\}.$$

The fundamental geometry may be seen as the complexified Riemannian geometry of the n-dimensional sphere S^n with the Euclidean inner product $\langle x, y \rangle$ replaced by the complex product $z \cdot w$. In particular the covariant derivative $\frac{D_z}{dt}(v)(t)$ of a curve of tangent vectors $v(t)$ along a curve $z(t) \in S_c^n$ is given by the formula

$$\frac{D_z}{dt}(v)(t) = \frac{dv}{dt} + (v(t) \cdot \frac{dz}{dt}(t))z(t).$$

The complex curvature $\kappa(t)$ of a curve whose tangent vector $\frac{dz}{dt}(t)$ satisfies $\frac{dz}{dt}(t) \cdot \frac{dz}{dt}(t) = 1$ is defined by

$$\kappa^2(t) = \frac{D_z}{dt}(\frac{dz}{dt})(t) \cdot \frac{D_z}{dt}(\frac{dz}{dt})(t).$$

Since only κ^2 figures in the calculations below it will not be necessary to go into any details concerning square roots of complex curves. Analogous to Definition 2.2, the distribution defined by the left invariant vector fields $X(g)$ on $SO_{n+1}(\mathbb{C})$ that take values in $\mathfrak{p}$ will be called horizontal, and its integral curves will be called horizontal. Any curve $z(t)$ on S_c^n can be lifted to a horizontal curve $g(t)$. If $g(t)$ is a solution of $\frac{dg}{dt}(t) = g(t)A(t)$ then $\frac{dz}{dt} \cdot \frac{dz}{dt} = \langle A(t), A(t) \rangle$ as can be seen through the argument used in Proposition 2.2. Then curves $V(t) = g(t)B(t)$ with $B(t) \in \mathfrak{p}$ can be interpreted as the curves of vectors along $g(t)$. The projection $\pi_*(\frac{D_g}{dt}(V))$ of the covariant derivative $\frac{D_g}{dt}(V) = g(t)\frac{dV}{dt}(t)$ satisfies

(55)
$$\pi_*(\frac{D_g}{dt}(V)) = \frac{D_{\pi(g)}}{dt}(\pi_*(g(V)).$$

Let us now return to the Hamiltonian equations of the Euler-Griffiths problem. The extremal curves $g(t)$ in $SO_{n+1}(\mathbb{C})$ are the solutions of

$$\frac{dg}{dt}(t) = g(t)(E_1 + \Omega(t)).$$

Each such curve defines a horizontal curve $h(t)$ defined by $h(t) = g(t)\phi^{-1}(t)$ where $\phi(t)$ is the solution of $\frac{d\phi}{dt}(t) = \phi(t)\Omega(t)$ with $\phi(0) = I$. It follows that $h(t)$ satisfies

$$\frac{dh}{dt}(t) = h(t)(\phi(t)E_1\phi^{-1}(t)).$$

Both $g(t)$ and $h(t)$ project onto the same curve $z(t) = g(t)e_0$ whose tangent vector $\frac{dz}{dt}(t)$ is given by $\frac{dz}{dt}(t) = g(t)e_1$. Therefore, $\frac{dz}{dt}(t) \cdot \frac{dz}{dt}(t) = 1$, and

$$\kappa^2(t) = \langle [E_1, \Omega(t)], [E_1, \Omega(t)] \rangle = \langle \Omega(t), \Omega(t) \rangle$$

as can be easily verified from equation (55) and the fact that $[[E_1, \Omega], E_1] = \Omega$.

PROPOSITION 7.1 (GRIFFITHS' THEOREM-PROBLEM 2).

(a) The complex function $\xi(t) = \kappa^2(t)$ associated with the elastic curve $z(t) = h(t)e_0$ satisfies the following differential equation:

(56)
$$\frac{d\xi}{dt}^2 = -\xi^3 + 4(\mathcal{H} - \epsilon)\xi^2 + 4(I_2 - \mathcal{H}^2)\xi + 4I_0$$

where $I_0 = -\langle P, P \rangle \langle \Omega, \Omega \rangle + \langle [\Omega, P], [\Omega, P] \rangle$ and $I_2 = \langle P, P \rangle + \epsilon \langle \Omega, \Omega \rangle$ are the integrals of motion.

(b) If $\tau(t)$ denotes the complex torsion associated with $z(t)$, then $(\xi(t)\tau(t))^2 = -I_0$. Moreover, the curve $z(t)$ is confined to a 3-dimensional complex sub- manifold of S_c^n.

PROOF. Since the proof pertains to Euler-Griffiths problem 2 where $\Delta = 0$, $M = \Omega$. Therefore, $I_0 = -\|\Omega\|^2\|P\|^2 + \|[\Omega, P]\|^2$ is an integral of motion for the above flow, as can be either verified directly, or it can be deduced from the discussion of the conservation laws in the previous section because $\Omega = u \wedge v$ with $u = \begin{pmatrix} 0 \\ M_1 \\ \vdots \\ M_{n-1} \end{pmatrix}$ and $v = e_0$. Furthermore, $I_2 = \|P\|^2 + \epsilon\|\Omega\|^2 = \|P\|^2 + \epsilon|M|^2$ along the flow, and therefore, it constitutes another integral of motion.

Consider now the complex curvature

$$\kappa^2(t) = \langle \Omega(t), \Omega(t) \rangle = \sum_{i=1}^{n-1} M_i^2(t)$$

traced by the above flow. Then,

$$\frac{d}{dt}(\kappa^2(t)) = 2\langle \Omega(t), [E_1, P(t)] \rangle = -2\sum_{i=1}^{n-1} M_i(t)p_{i+1}(t),$$

and therefore,

$$(\frac{d}{dt}(\xi(t))^2 = 4(\sum_{i=1}^{n-1} M_i p_{i+1})^2 = 4(\|\Omega\hat{P}\|^2 - p_1^2\|\Omega\|^2)$$

$$= 4(I_0 - \|\Omega\|^2 p_1^2 + \|P\|^2\|\Omega\|^2)$$

$$= 4(I_0 - \kappa^2(\mathcal{H} - \frac{1}{2}\kappa^2)^2 + (I_2 - \epsilon\|\Omega\|^2)\|\Omega\|^2)$$

$$= 4(I_0 - \kappa^2(\mathcal{H} - \frac{1}{2}\kappa^2)^2 + (I_2 - \epsilon\kappa^2)\kappa^2)$$

$$- \xi^3(t) + 4(\mathcal{H} - \epsilon)\xi^2(t) + 4(I_2 - \mathcal{H}^2)\xi(t) + 4I_0$$

This equation is the complex extension of the equation obtained by P. Griffiths with $\mathcal{H} = 0$ and $\epsilon = 0$ ([**Gr**],p. 153).

To show part (b) $T(t) = \phi(t)E_1\phi^{-1}(t)$. Then

$$\frac{dT}{dt}(t) = \phi(t)[E_1, \Omega(t)]\phi^{-1}(t) = \kappa(t)N(t).$$

The Serret-Frenet relations define the torsion $\tau(t)$ in terms of the triad T, N, B given by the following differential equations:

$$\frac{dN}{dt}(t) = -\kappa(t)T(t) + \tau(t)B(t) \text{ and } \frac{dB}{dt}(t) = -\tau(t)N(t)$$

Therefore,

$$\frac{d^2T}{dt^2}(t) = \phi(t)[[E_1, \Omega(t)], \Omega(t)]\phi^{-1}(t) + \phi(t)[E_1, \frac{d\Omega}{dt}(t)]\phi^{-1}(t)$$

$$= \frac{d\kappa}{dt}(t)N(t) + \kappa(t)\frac{dN}{dt}(t)$$

$$= \frac{d\kappa}{dt}(t)N(t) - \kappa^2(t)T(t) + \kappa(t)\tau(t)B(t).$$

It follows by an easy computation that

$$[[E_1, \Omega], \Omega] = \langle -\Omega, \Omega \rangle E_1 \text{ and } [E_1, \frac{d\Omega}{dt}] = [E_1, [E_1, P]] = -P + p_1 E_1$$

hence,

$$\phi(t)(-P(t) + p_1(t)E_1)\phi^{-1} = \frac{d\kappa}{dt}(t)N(t) + \kappa(t)\tau(t)B(t).$$

But then

$$(\kappa^2(t)\tau(t))^2 =$$

$$= \langle (\kappa(t)(-P(t) + p_1(t)E_1) - \kappa(t)\frac{d\kappa}{dt}(t)N(t), (\kappa(t)(-P(t) + p_1(t)E_1) - \kappa(t)\frac{d\kappa}{dt}(t)N(t) \rangle$$

$$= \kappa^2(t)(\langle P(t), P(t) \rangle - p_1^2(t)) + 2\kappa(t)\frac{d\kappa}{dt}(t)\langle P(t), [E_1, \Omega(t)] \rangle + (\kappa(t)\frac{d\kappa}{dt}(t))^2$$

$$= \kappa^2(t)(\langle P(t), P(t) \rangle - p_1^2(t)) - 2(\kappa(t)\frac{d\kappa}{dt}(t))^2 + (\kappa(t)\frac{d\kappa}{dt}(t))^2$$

$$= \kappa^2(t)(\langle P(t), P(t) \rangle - p_1^2(t)) + (\kappa(t)\frac{d\kappa}{dt}(t))^2$$

$$= \langle \Omega(t), \Omega(t) \rangle \langle P(t), P(t) \rangle - (\langle \Omega(t), \Omega(t) \rangle p_1^2(t) + (\sum_{i=1}^{n-1} M_i(t)p_{i+1}(t))^2)$$

$$= \langle \Omega(t), \Omega(t) \rangle \langle P(t), P(t) \rangle - \langle [P(t), \Omega(t)], [P(t), \Omega(t)] \rangle = -I_0.$$

We leave it to the reader to verify that $\frac{dB}{dt}(t)$ is contained in the linear span of $T(t), N(t), B(t)$.

The above proposition provides a complete solution to the Euler-Griffiths problem 2: the solution of equation (56) defines the curvature of the elastic curve associated with an extremal curve. Once the curvature is known, the torsion is also known since $\kappa^2\tau$ is constant. Then the solution of the Serret-Frenet equations determines the corresponding elastic curve $z(t)$.

The solutions to Euler-Griffiths problem 1 are more complicated. To begin with, the solutions to problem 1 do not necessarily conform to the condition that $\kappa^2\tau$ is constant. Indeed, the preceeding equations readily imply that $\kappa^2\tau$ is constant if and only if $\Delta = 0$. Furthermore, there does not seem to be a clear path to the differential equation for $\xi = \kappa^2$ in dimensions greater than 3. For these reasons we shall not go further into this problem.

Instead we shall finish this section with an aside remark pertaining to the solutions of the Hamiltonian $\mathcal{H}$ equal to

$$\mathcal{H} = \langle P, P \rangle.$$

Such a Hamiltonian is the simplest among the Hamiltonians that are quadratic relative to the Cartan space $\mathfrak{p}$. This Hamiltonian may be seen as the complexifications of the Hamiltonian for the standard geodesic problem on the space form M_ϵ. Like the Euler-Griffiths problem, $\mathcal{H}$ also has a large symmetry group, namely $K = \{1\} \times SO_n(\mathbb{C})$. Consequently, the integral curves $g(t)$ are of the form

$$g(t) = e^{t(A+B)} e^{-tB}$$

for arbitrary matrices $A \in \mathfrak{p}$ and $B \in \mathfrak{k}$. The complex geodesics $z(t)$ on S_c^n are of the form

(57) $$z(t) = g_0 e^{tA} e_0,$$

because of the transversality conditions ($B = 0$, similar to the Euler-Griffiths problem 1). Equation (57) when restricted to the real forms of $SL_2(\mathbb{C})$ yields the geodesics for three dimensional spaces of constant curvature. In particular, for $SO_3(\mathbb{R})$ the curves in (57) trace great circles on S^3, while for $SL_2(\mathbb{R})$ they trace hyperbolic geodesics in the Poincaré upper half plane. The other real forms $SU(1,1)$ and $SO(1,2)$ correspond to other isometric models of hyperbolic geometry, namely the unit disc D and the hyperboloid $\mathbb{H}^2$.

VII. Kirchhoff's elastic Problems on six dimensional Lie Groups

This chapter, a central part of the paper, may be viewed as a synthesis of the theory of tops extended to complex Hamiltonians via the elastic problems. The investigations of the integrable cases begin with a classification of systems having meromorphic solutions, an idea that originated with S. Kowalewski in her brilliant, but somewhat opaque paper on the top published in 1889 ([**Ko**]). As will be shown below her classification of integrable tops carries over to the elastic problems of Kirchhoff, and as in the case of the top, identifies likely candidates that admit an extra integral of motion resulting in a completely integrable Hamiltonian system.

1. Meromorphic solutions.

The most general Hamiltonian $\mathcal{H}$ that corresponds to an elastic problem of Kirchhoff is of the form

$$\mathcal{H} = \frac{1}{2}\left(\frac{M_1^2}{\lambda_1} + \frac{M_2^2}{\lambda_2} + \frac{M_3^2}{\lambda_3}\right) + b_1 p_1 + b_2 p_2 + b_3 p_3$$

where the variables $M_1, M_2, M_3, p_1, p_2, p_3$ are the coordinates of $\ell \in \mathfrak{g}^*$ relative to the dual basis of $A_1, A_2, \Lambda_3, B_1, B_2, B_3$ defined in Section IV. Our immediate objective is to seek conditions on the parameters λ_1, λ_2, λ_3, b_1, b_2, b_3 so that the integral curves of $\vec{\mathcal{H}}$ are meromorphic functions of complex time. To cover all three real space forms we assume that λ_1, λ_2, λ_3 are real and that b_1, b_2, b_3 are either all real or all imaginary. The imaginary choice covers the hyperbolic case.

We recall the Hamiltonian equations of $\mathcal{H}$:

$$\frac{dM}{dt} = [\Omega, M] + [B, P] \quad \frac{dP}{dt} = [\Omega, P] + \epsilon[B, M],$$

which can be also written as

(58)
$$\frac{d\hat{M}}{dt} = \hat{M} \times \hat{\Omega} + \hat{P} \times \hat{B}\ , \quad \frac{d\hat{P}}{dt} = \hat{P} \times \hat{\Omega} + \epsilon(\hat{M} \times B)$$

with $\epsilon = 0, \pm 1$ and where each vector

$$\hat{A} = \begin{pmatrix} a_1 \\ a_2 \\ a_3 \end{pmatrix} \text{ corresponds to the matrix } A = \begin{pmatrix} 0 & -a_3 & a_2 \\ a_3 & 0 & -a_1 \\ -a_2 & a_1 & 0 \end{pmatrix}.$$

DEFINITION 7.1. *The preceding system of equations will be referred to as a meromorphic case (relative to the parameters λ_1, λ_2, λ_3, b_1, b_2, b_3) if all the solutions are meromorphic functions of complex time.*

The most immediate aim is to classify the meromorphic cases.

My earlier paper [**Ju 3**] classified the meromorphic cases, under the assumption that two eigenvalues are equal and under an additional assumption that certain non-degeneracy condition must hold. More precisely, the paper assumed that the general meromorphic solution is of the form

$$M(t) = t^{-n_1}(M_0 + M_1 t + \cdots)\ , \quad P(t) = t^{-n_2}(P_0 + P_1 t + \cdots)$$

where $[M_0, P_0] \neq 0$. The assumption that $[M_0, P_0] \neq 0$ implies that $n_1 = 1$ and $n_2 = 2$. In her original paper of 1889, S. Kowalewski implicitly assumed that

$$M(t) = \frac{1}{t}M_0 + M_1 + M_2 t + \cdots \quad \text{and that}$$

$$P(t) = \frac{1}{t^2}P_0 + \frac{1}{t}P_1 + P_2 + P_3 t + \cdots,$$

and then proceeded to classify the meromorphic cases for the top.

This gap in the paper of Kowalewski was first noticed by A.A. Markov, and subsequently there appeared several publications in Russia which attempted to provide the correct proofs for the claim of Kowalewski. The issue was finally settled by A.M. Lyapunov in 1894 ([**Ly**]).

In this paper, Lyapunov considered the tangent map induced by the flow of equation (58) for $\epsilon = 0$, and then argued that the tangent map must be single valued along each point of the flow whenever the flow is meromorphic. Alternatively stated, the solutions of the variational equation associated with each trajectory must be single valued. He then based his analysis on the variational equation associated with the solution of the form

$$M(t) = \frac{1}{t}M_0 \quad \text{and} \quad P(t) = \frac{1}{t^2}P_0$$

for some constants M_0 and P_0.

The basic arguments of Lyapunov can be adapted to the problems of Kirchhoff for the following reasons. The parameter ϵ may be considered as a real parameter, in which case it will be shown that the meromorphic cases with $\epsilon = 0$, the case considered by Lyapunov, coincide with the meromorphic cases for any number ϵ.

The following lemma is basic.

LEMMA 1. *Let $P(t)$ and $M(t)$ be any solutions of (58), and let μ be any non-zero number. Then $M(t)$ and $Q(t) = \frac{1}{\mu}P(t)$ are the solutions of*

$$\frac{dM}{dt} = [\Omega, M] + [B_\mu, Q] , \quad \frac{dQ}{dt} = [\Omega, Q] + \frac{\epsilon}{\mu^2}[B_\mu, M]$$

where $B_\mu = \mu B$.

PROOF. Evidently $[B_\mu, Q] = [B, P]$, and

$$\frac{dQ}{dt} = \frac{1}{\mu}\frac{dP}{dt} = \frac{1}{\mu}\left([\Omega, P] + \epsilon[B, M]\right) = [\Omega, Q] + \frac{\epsilon}{\mu}[B, M] = [\Omega, Q] + \frac{\epsilon}{\mu^2}[B_\mu, M] \ .$$

It is a corollary of the preceeding lemma that the meromorphic cases are invariant under the dilations $P \to \frac{1}{\mu}P(t)$ and $B \to \mu B$ when $\epsilon = 0$. Therefore, the meromorphic cases depend only on the direction of B. Secondly, the lemma implies that for $\epsilon \neq 0$ it is sufficient to classify meromorphic cases for small ϵ.

THEOREM 2. Let $\epsilon = 0$ and $B \neq 0$. Then meromorphic solutions require that at least two eigenvalues λ_1, λ_2, λ_3 be equal to each other.

PROOF. The proof of this theorem will proceed via several lemmas. Assume first that λ_1, λ_2, λ_3 are all distinct.

LEMMA 2. Meromorphic solutions require that one of the coordinates of $\hat{B}$ be zero.

PROOF. Introduce the following constants

$$a = \sqrt{\frac{\lambda_1}{\lambda_3 - \lambda_2}} \ , \quad b = \sqrt{\frac{\lambda_2}{\lambda_1 - \lambda_3}} \ , \quad c = \sqrt{\frac{\lambda_3}{\lambda_2 - \lambda_1}}$$

and consider $\hat{M}(t) = \frac{1}{t}M_0$, where $M_0 = \begin{pmatrix} bc\lambda_1 \\ ac\lambda_2 \\ ab\lambda_3 \end{pmatrix}$. Then

$$\hat{\Omega} = \frac{1}{t}\begin{pmatrix} \frac{1}{\lambda_1}bc\lambda_1 \\ \frac{1}{\lambda_2}ac\lambda_2 \\ \frac{1}{\lambda_3}ab\lambda_3 \end{pmatrix} = \frac{1}{t}\begin{pmatrix} bc \\ ac \\ ab \end{pmatrix}, \text{ and}$$

$$\hat{M}(t) \times \hat{\Omega}(t) = -\frac{1}{t^2}M_0 = \frac{d\hat{M}}{dt}(t).$$

It follows that $\hat{M}(t)$ is a solution of $\frac{d\hat{M}}{dt} = \hat{M} \times \hat{\Omega}$ and therefore, $\hat{M}(t)$ together with $\hat{P}(t) = 0$ is a solution of equation (58) with $\epsilon = 0$. We now consider the variational equations associated this trajectory.

First recall the basic facts about the variational equations. Let $u(\tau)$ and $v(\tau)$ denote arbitrary differentiable curves that satisfy $u(0) = \hat{M}(1)$ and $v(0) = \hat{P}(1)$, and $\hat{M}_\tau(t)$ and $\hat{P}_\tau(t)$ denote the solutions of (58) that pass through $u(\tau)$ and $v(\tau)$ at $t = 1$. Let $U(t) = \frac{d}{d\tau}\hat{M}_\tau(t)\big|_{\tau=0}$

and $V(t) = \frac{d}{d\tau}\hat{P}_\tau(t)\big|_{\tau=0}$. In addition let $Q(t) = \begin{pmatrix} \frac{1}{\lambda_1}u_1(t) \\ \frac{1}{\lambda_2}u_2(t) \\ \frac{1}{\lambda_3}u_3(t) \end{pmatrix}$ where $u_1(t), u_2(t), u_3(t)$ denotes the coordinates of $U(t)$. Then $U(t)$ and $V(t)$ are the solutions of the variational equation

(59)
$$\frac{dU}{dt} = U \times \hat{\Omega} + \hat{M} \times Q + V \times \hat{B} \text{ and}$$

(60)
$$\frac{dV}{dt} = V \times \hat{\Omega}$$

with the initial data $U(1) = \frac{du}{d\tau}(0)$ and $V(1) = \frac{dv}{d\tau}(0)$.

Equation (60) is of the form $\frac{dV}{dt} = -\frac{1}{t}\Lambda V$ where

$$\Lambda = \begin{pmatrix} 0 & -ab & ac \\ ab & 0 & -bc \\ -ac & bc & 0 \end{pmatrix}$$

and equation (59) is of the form

$$\frac{dU}{dt}(t) = A(t)U(t) + V(t) \times \hat{B}$$

where

$$A(t)U(t) = U(t) \times \hat{\Omega}(t) + \hat{M}(t) \times Q(t).$$

An easy calculation which we omit shows that $A(t) = \frac{1}{t}\Delta$ with

$$\Delta = \begin{pmatrix} 0 & ab\lambda_3\left(\frac{1}{\lambda_3}-\frac{1}{\lambda_2}\right) & ac\lambda_2\left(\frac{1}{\lambda_3}-\frac{1}{\lambda_2}\right) \\ ab\lambda_3\left(\frac{1}{\lambda_1}-\frac{1}{\lambda_3}\right) & 0 & bc\lambda_1\left(\frac{1}{\lambda_1}-\frac{1}{\lambda_3}\right) \\ ac\lambda_2\left(\frac{1}{\lambda_2}-\frac{1}{\lambda_1}\right) & bc\lambda_1\left(\frac{1}{\lambda_2}-\frac{1}{\lambda_1}\right) & 0 \end{pmatrix}.$$

We shall first investigate the solutions of $\frac{dV}{dt} = -\frac{1}{t}\Lambda V$. Note that

$$(ab)^2 + (ac)^2 + (bc)^2$$
$$= \frac{\lambda_1\lambda_2}{(\lambda_3-\lambda_2)(\lambda_1-\lambda_3)} + \frac{\lambda_1\lambda_3}{(\lambda_3-\lambda_2)(\lambda_2-\lambda_1)} + \frac{\lambda_2\lambda_3}{(\lambda_1-\lambda_3)(\lambda_2-\lambda_1)}$$
$$= \frac{\lambda_1\lambda_2(\lambda_2-\lambda_1) + \lambda_1\lambda_3(\lambda_1-\lambda_3) + \lambda_2\lambda_3(\lambda_3-\lambda_2)}{(\lambda_3-\lambda_2)(\lambda_1-\lambda_3)(\lambda_2-\lambda_1)} = -1.$$

It follows that the characteristic polynomial of Λ is equal to $\xi(\xi^2 + ((ab)^2 + (ac)^2 + (bc)^2)) = \xi(\xi^2 - 1)$, and therefore, the eigenvalues of Λ are

$$\xi_1 = 1 \ , \ \xi_2 = -1 \ , \ \xi_3 = 0 \ .$$

Then $\mathbb{C}^3 = W_1 \oplus W_2 \oplus W_3$ where W_1, W_2, W_3 denote the eigenspaces of Λ. It is easy to verify that

$$W_1 = \{w \in \mathbb{C}^3 : \hat{\Lambda} \times w = w\}, \ W_2 = \{w \in \mathbb{C}^3 : \hat{\Lambda} \times w = -w\} \text{ and } W_3 = \mathbb{C}\hat{\Lambda}$$

from which it follows that the eigenspaces W_1 and W_2 are isotropic, in the sense that their vectors satisfy $w \cdot w = 0$. These eigenvectors also satisfy $\hat{\Lambda} \cdot w = 0$.

Equation (60) becomes

$$\dot{w}_1(t) = -\frac{1}{t}w_1(t) \ , \ \dot{w}_2(t) = \frac{1}{t}w_2(t) \ , \ \dot{w}_3(t) = 0$$

whenever $V(t)$ is written as $V(t) = w_1(t) + w_2(t) + w_3(t)$ with $w_i(t) \in W_i$, $i = 1, 2, 3$. Hence,

(61) $$w_1(t) = \frac{1}{t}w_1(1) \ , \ w_2(t) = tw_2(1) \ , \ w_3(t) = \text{ constant } .$$

To calculate the eigenvalues of Δ it will be convenient to introduce the following notations:

$$\delta_1 = bc\lambda_1\left(\frac{1}{\lambda_1} - \frac{1}{\lambda_3}\right) \ , \ \delta_2 = ac\lambda_2\left(\frac{1}{\lambda_3} - \frac{1}{\lambda_2}\right) \ , \ \delta_3 = ab\lambda_3\left(\frac{1}{\lambda_3} - \frac{1}{\lambda_2}\right) .$$

It follows that

$$\delta_1 = bc\lambda_1\left(\frac{1}{\lambda_1} - \frac{1}{\lambda_3}\right) = \sqrt{\frac{\lambda_2}{\lambda_1 - \lambda_3}}\sqrt{\frac{\lambda_3}{\lambda_2 - \lambda_1}}\frac{\lambda_3 - \lambda_1}{\lambda_3} = -\sqrt{\left(\frac{\lambda_1 - \lambda_3}{\lambda_2 - \lambda_1}\right)}\frac{\lambda_2}{\lambda_3} = -\frac{c\lambda_2}{b\lambda_3}$$

and $bc\lambda_1\left(\frac{1}{\lambda_2} - \frac{1}{\lambda_1}\right) = \frac{1}{\delta_1}.$

Similar calculations show that

$$\delta_2 = -\frac{c\lambda_1}{a\lambda_3}, \ ac\lambda_2\left(\frac{1}{\lambda_2} - \frac{1}{\lambda_1}\right) = \frac{1}{\delta_2} \ , \quad \text{and} \quad \delta_3 = -\frac{b\lambda_1}{a\lambda_2}, \ ab\lambda_3\left(\frac{1}{\lambda_1} - \frac{1}{\lambda_3}\right) = \frac{1}{\delta_3} .$$

Therefore,

$$\Delta = \begin{pmatrix} 0 & \delta_3 & \delta_2 \\ \dfrac{1}{\delta_3} & 0 & \delta_1 \\ \dfrac{1}{\delta_2} & \dfrac{1}{\delta_1} & 0 \end{pmatrix} .$$

Furthermore,

$$\frac{\delta_2}{\delta_1\delta_3} = \left(-\frac{A}{B}\frac{\lambda_2}{\lambda}\right)\left(-\frac{B}{C}\frac{\lambda_3}{\lambda_2}\right)\left(-\frac{C}{A}\frac{\lambda_1}{\lambda_3}\right) = -1$$

which in turn implies that $\det \Delta = -2$.

Then the characteristic polynomial of Δ is equal to $-\xi^3 + 3\xi - 2 = -(1 - \xi)^2(\xi + 2)$, and therefore Δ has a double eigenvalue $\xi_{1,2} = 1$ and a single eigenvalue $\xi_3 = -2$.

An easy calculation shows that the eigenspace V_1 corresponding to $\xi = 1$ is two dimensional given by $V_1 = \{x \in \mathbb{C}^3 : x_1 = \delta_3 x_2 + \delta_2 x_3\}$. The normal vector to V_1 is a scalar multiple of $(1, -\delta_3, -\delta_2)$. It follows that the direction of the normal is given by

$$n = (\frac{a}{\lambda_1}, \frac{b}{\lambda_2}, \frac{c}{\lambda_3}).$$

The eigenspace V_{-2} corresponding to $\xi_3 = -2$ is given by $V_{-2} = \{x \in \mathbb{C}^3 : x_1 = -\delta_3 x_2, x_2 = -\delta_1 x_3\}$, which is the complex line through the point

$$t = (\frac{\lambda_1}{a}, \frac{\lambda_2}{b}, \frac{\lambda_3}{c}).$$

It then follows that the general solution of $\dfrac{dx}{dt} = \dfrac{1}{t}\Delta x(t)$ is given by the sum of solutions

$$x(t) = tx(1) \quad \text{with} \quad x(1) \in V_1 , \quad \text{and}$$
$$x(t) = \frac{1}{t^2}x(1) \quad \text{with} \quad x(1) \in V_{-2} .$$

Equation (59) is of the form

$$\frac{dx}{dt} = \frac{1}{t}\Delta x(t) + f(t) \quad \text{where} \quad f(t) = V(t) \times \hat{B}.$$

Let $V(t) = w_3$ for an arbitrary point $w_3 \in W_3$. Since $(bc)^2 + (ac)^2 + (ab)^2 = -1$, W_3 cannot contain any real vectors. In particular $\hat{B}$ is not colinear with any w_3 in W_3, and hence $f = w_3 \times \hat{B}$ is not zero.

Let $f = f_1 + f_{-2}$ with $f_1 \in V_1$ and f_{-2} in V_{-2}. If $f_1 \neq 0$ then the solutions of (59) cannot be all single valued for the following reason:

The restriction of (59) to the eigenspace V_1 is of the form

$$\frac{dx}{dt} = \frac{1}{t}x(t) + f_1 .$$

A particular solution $x_p(t)$ is of the form $x_p(t) = t\varphi(t)$ for some function $\varphi(t)$. But then $\dfrac{dx_p}{dt} = t\varphi' + \varphi(t) = \dfrac{1}{t}x_p + f_1$, and therefore, $\varphi'(t) = \dfrac{f_1}{t}$. But then φ is multivalued, and hence x_p is multivalued as well.

Therefore meromorphic solutions require that f belongs to V_{-2}. This condition implies that

$$b_2 ab - b_3 ac = t\frac{\lambda_1}{a}$$
$$b_3 bc - b_1 ab = t\frac{\lambda_2}{b}$$
$$b_1 ac - b_2 bc = t\frac{\lambda_3}{c}$$

for some complex number t. It follows that $\frac{b_2 b - b_3 c}{b_3 c - b_1 a} = \frac{\lambda_3 - \lambda_2}{\lambda_1 - \lambda_3}$, which in turn implies that

$$ab_1(\lambda_3 - \lambda_2) + bb_2(\lambda_1 - \lambda_3) + cb_3(\lambda_2 - \lambda_1) = 0.$$

Therefore,

$$b_1\sqrt{\lambda_1(\lambda_3 - \lambda_2)} + b_2\sqrt{\lambda_2(\lambda_1 - \lambda_3)} + b_3\sqrt{\lambda_3(\lambda_2 - \lambda_1)} = 0.$$

Either one of the above roots is imaginary and the other two roots are real, or the opposite is true. In either case one of b_1, b_2, b_3 must be zero. $\square$

Until the end of the proof of Theorem 2 it will be assumed that $\lambda_1 > \lambda_2 > \lambda_3$ in which case $b_2 = 0$, and

(62)
$$b_1\sqrt{\lambda_1(\lambda_3 - \lambda_2)} + b_3\sqrt{\lambda_3(\lambda_2 - \lambda_1)} = 0.$$

LEMMA 3.

$$\hat{M}(t) = \frac{1}{t}\begin{pmatrix} 0 \\ 2i\lambda_2 \\ 0 \end{pmatrix} \quad and \quad \hat{P}(t) = \frac{1}{t^2}\begin{pmatrix} ih \\ 0 \\ h \end{pmatrix}$$

is a solution of equation (58) where $h = \frac{2\lambda_2}{b}$ and $b = b_3 + ib_1$.

The verification of this lemma will be left to the reader.

Consider now the variational equation associated with the above flow. It is obtained in a manner similar to the previous case, and we have

$$\frac{dU}{dt} = U \times \frac{1}{t}\begin{pmatrix} 0 \\ 2i \\ 0 \end{pmatrix} + \frac{1}{t}\begin{pmatrix} 0 \\ 2i\lambda_2 \\ 0 \end{pmatrix} \times \begin{pmatrix} \frac{1}{\lambda_1}u_1 \\ \frac{1}{\lambda_2}u_2 \\ \frac{1}{\lambda_3}u_3 \end{pmatrix} + V \times \hat{B},$$

$$\frac{dV}{dt} = V \times \frac{1}{t}\begin{pmatrix} 0 \\ 2i \\ 0 \end{pmatrix} + \hat{P} \times \begin{pmatrix} \frac{1}{\lambda_1}u_1 \\ \frac{1}{\lambda_2}u_2 \\ \frac{1}{\lambda_3}u_3 \end{pmatrix}$$

If $W(t) = tV(t)$, then the preceeding equations become

$$\frac{dU}{dt} = \frac{1}{t}[U \times \begin{pmatrix} 0 \\ 2i \\ 0 \end{pmatrix} + \begin{pmatrix} 0 \\ 2i\lambda_2 \\ 0 \end{pmatrix} \times \begin{pmatrix} \frac{1}{\lambda_1}u_1 \\ \frac{1}{\lambda_2}u_2 \\ \frac{1}{\lambda_3}u_3 \end{pmatrix} + W \times \hat{B}],$$

$$\frac{dW}{dt} = \frac{1}{t}[W + W \times \begin{pmatrix} 0 \\ 2i \\ 0 \end{pmatrix} + \begin{pmatrix} ih \\ 0 \\ h \end{pmatrix} \times \begin{pmatrix} \frac{1}{\lambda_1}u_1 \\ \frac{1}{\lambda_2}u_2 \\ \frac{1}{\lambda_3}u_3 \end{pmatrix}]$$

The above differential system breaks up into two independent subsystems:

$$\frac{du_1}{dt} = \frac{1}{t}(-2iu_3 + 2i\frac{\lambda_2}{\lambda_3}u_3 + w_2 b_3)$$

$$\frac{du_3}{dt} = \frac{1}{t}(2iu_1 - 2i\frac{\lambda_2}{\lambda_1}u_1 - w_2 b_1)$$

$$\frac{dw_2}{dt} = \frac{1}{t}(w_2 + h\frac{u_1}{\lambda_1} - ih\frac{u_3}{\lambda_3}),$$

and

$$\frac{du_2}{dt} = \frac{1}{t}(w_3 b_1 - b_3 w_1)$$

$$\frac{dw_1}{dt} = \frac{1}{t}(w_1 + 2iw_3 - \frac{h}{\lambda_2}u_2)$$

$$\frac{dw_3}{dt} = \frac{1}{t}(w_3 - 2iw_1 + i\frac{h}{\lambda_2}u_2)$$

which are then described by the matrices

$$A = \begin{pmatrix} 0 & -2i(\frac{-\lambda_3+\lambda_2}{\lambda_3}) & b_3 \\ 2i(\frac{\lambda_1-\lambda_2}{\lambda_1}) & 0 & -b_1 \\ h\frac{1}{\lambda_1} & -ih\frac{1}{\lambda_3} & 1 \end{pmatrix} \text{ and } B = \begin{pmatrix} 0 & -b_3 & b_1 \\ -\frac{h}{\lambda_2} & 1 & 2i \\ i\frac{h}{\lambda_2} & -2i & 1 \end{pmatrix}.$$

The solutions of these two subsystems must be single valued under the assumption that the integral curves of differential system (58) are meromorphic functions of t). But the solutions of the preceeding linear systems are single valued only when the eigenvalues of A and B are integers. Therefore, if either of the above matrices has an eigenvalue λ which is not an integer, then differential system (58) cannot have only meromorphic solutions.

The reader can readily verify that matrix B has eigenvalues $\lambda = 1, \lambda = 3$, and $\lambda = -2$. However, the determinant of A is complex unless $b = 0$ and therefore, the eigenvalues of A cannot be all real when $\hat{B} \neq 0$. We leave this calculation to the reader.

We have now shown that equation (58) does not have meromorphic solutions when $\epsilon = 0$, $\lambda_1, \lambda_2, \lambda_3$ are distinct and $\hat{B} \neq 0$. To show that the same conclusion holds for all ϵ we point out that the solutions of an analytic differential system with a parameter depend analytically on the parameter. Therefore each solution $(\hat{M}(t), \hat{P}(t))$ of (58) can be developed as

$$\hat{M}(t) = M_0(t) + \epsilon M_1(t) + \epsilon^2 M_2(t) + \cdots$$

and

$$\hat{P}(t) = P_0(t) + \epsilon P_1(t) + \epsilon^2 P_2(t) + \cdots$$

It follows that for small values of ϵ each $M_i(t)$ and $P_i(t)$ must be meromorphic if $(\hat{M}(t), \hat{P}(t))$ is to be meromorphic. Since $(M_0(t), P_0(t))$ is a solution of (58) with $\epsilon = 0$ it follows from above that $(\hat{M}(t), \hat{P}(t))$ cannot be meromorphic for small ϵ. But then the same must be true for all ϵ as a consequence of Lemma 1. The proof of Theorem 2 is now complete.   □

THEOREM 3. *Assume that $\hat{B} \neq 0$ and that $\lambda_1 = \lambda_2$. Then the solutions of equation (58) are meromorphic functions of complex time t only in the following cases:*

(10) *(i)* $\lambda_1 = \lambda_2 = \lambda_3$.

(11) *(ii)* $\lambda_1 = \lambda_2$, $b_1 = b_2 = 0$. *(Lagrange's case).*

(12) *(iii)* $\lambda_1 = \lambda_2 = 2\lambda_3$, $b_3 = 0$. *(Kowalewski's case).*

Note that under the rotations around the e_1-axis the drift vector $\hat{B}$ in equations (58) is rotated to a new position but that otherwise the system remains invariant. Therefore, there is no loss in generality in assuming that $\hat{B}$ of the form $(b_1, 0, b_3)$ with $b_1 \neq 0$. In addition to this assumption, assume at first that $\epsilon = 0$. Then

LEMMA 4. *The only solutions of the form*

$$\hat{M}(t) = \frac{1}{t}\begin{pmatrix} p \\ q \\ r \end{pmatrix}, \quad and \quad \hat{P}(t) = \frac{1}{t^2}\begin{pmatrix} f \\ g \\ h \end{pmatrix}.$$

are given by the following cases:

CASE 1. $p = r = 0$, $q = 2i\lambda_1$, $f = ih$, $g = 0$, $h = \dfrac{2\lambda_1}{b_3 + ib_1}$,

and

CASE 2. $p = iq$, $q(2\lambda_3 - \lambda_1) = \dfrac{2b_3}{b_1}\lambda_1\lambda_3$, $r = 2i\lambda_3$, $f = \dfrac{2\lambda_3}{b_1}$, $g = if$, $h = 0$.

PROOF. It follows from (58) that

$$-M_0 = [\Omega_0, M_0] + [B, P_0], \text{ and } -2P_0 = [\Omega_0, P_0]$$

where $\hat{\Omega}_0 = \begin{pmatrix} \dfrac{1}{\lambda_1}p_1 \\ \dfrac{1}{\lambda_2}q \\ \dfrac{1}{\lambda_3}r \end{pmatrix}$, $\hat{M}_0 = \begin{pmatrix} p \\ q \\ r \end{pmatrix}$ and $\hat{P}_0 = \begin{pmatrix} f \\ g \\ h \end{pmatrix}$.

Therefore,

$$-p = qr\left(\frac{1}{\lambda_3} - \frac{1}{\lambda_2}\right) + gb_3 \quad , \quad -2f = \frac{gr}{\lambda_3} - \frac{hq}{\lambda_2}$$

$$-q = pr\left(\frac{1}{\lambda_1} - \frac{1}{\lambda_3}\right) + hb_1 - fb_3 \quad , \quad -2g = \frac{hp}{\lambda_1} - \frac{fr}{\lambda_3}$$

$$-r = pq\left(\frac{1}{\lambda_2} - \frac{1}{\lambda_1}\right) - gb_1 \quad , \quad -2h = \frac{fq}{\lambda_2} - \frac{gp}{\lambda_1}.$$

The fact that both $\hat{M}(t) \cdot \hat{P}(t)$ and $\hat{P}(t) \cdot \hat{P}(t)$ are constant along the solutions of (58) implies that $pf + qg + rh = 0$ and $f^2 + g^2 + h^2 = 0$. In addition $\hat{P}_0 \cdot \hat{\Omega}_0 = 0$ which further implies that $\dfrac{1}{\lambda_1}pf + \dfrac{1}{\lambda_2}qg + \dfrac{1}{\lambda_3}rh = 0$.

Since $\lambda_1 = \lambda_2$ it follows from above that $rh = 0$.

Assume that $r = 0$ (Case 1). Then $g = 0$, since b_1 is assumed nonzero, and therefore, $p = 0$. Finally, $f = ih$, $q = fb_3 - hb_1 = ih(b_3 + ib_1)$, and $-2ih = -\dfrac{hq}{\lambda_2}$. It follows that $q = 2i\lambda_2 = 2i\lambda_1$ and

$$h = \frac{2\lambda_1}{b_3 + ib_1}.$$

The calculations in the case $h = 0$ are similar and will be omitted.

□

Consider now the variational equations associated the trajectories in the preceeding lemma. Let $z(\varepsilon)$ and $w(\epsilon)$ denote arbitrary differentiable curves that satisfy $z(0) = M(1) = M_0$ and $w(0) = P(1) = P_0$. Denote by u and v the tangents of z and w at zero. Let $M_\varepsilon(t)$ and $P_\varepsilon(t)$ denote the solutions of (58) passing through $z(\epsilon)$ and $w(\epsilon)$ at $t = 1$, and let $U(t) = \dfrac{d}{d\varepsilon}M_\varepsilon(t)\big|_{\varepsilon=0}$, and let $V(t) = \dfrac{d}{d\varepsilon}P_\varepsilon(t)\big|_{\varepsilon=0}$.

Then $U(t)$ and $V(t)$ are the solutions of the following equations

$$\frac{dU}{dt} = [\Omega, U] + [\Lambda, M] + [B, V], \text{ and}$$

$$\frac{dV}{dt} = [\Omega, V] + [\Lambda, P]$$

where $\hat{U}(t) = \begin{pmatrix} u_1(t) \\ u_2(t) \\ u_3(t) \end{pmatrix}$ and $\hat{\Lambda}(t) = \begin{pmatrix} \dfrac{1}{\lambda_1}u_1(t) \\ \dfrac{1}{\lambda_2}u_2(t) \\ \dfrac{1}{\lambda_3}u_3(t) \end{pmatrix}$.

Since $M(t) = \dfrac{1}{t} M_0$, $\Omega(t) = \frac{1}{t}\Omega_0$ and $P(t) = \dfrac{1}{t^2} P_0$,

$$\frac{dU}{dt} = \frac{1}{t}([\Omega_0, U] + [\Lambda, M_0]) + [B, V]$$
$$\frac{dV}{dt} = \frac{1}{t}[\Omega_0, V] + \frac{1}{t^2}[\Lambda, P_0]$$

Define a new variable $W = tV(t)$. Then,

$$\frac{dU}{dt} = \frac{1}{t}([\Omega_0, U] + [\Lambda, M_0] + [B, W]) , \quad \text{and}$$
$$\frac{dV}{dt} = \frac{1}{t}(W + [\Omega_0, W] + [\Lambda, P_0]).$$

Following Lyapunov's arguments, we conclude that $U(t)$ and $W(t)$ will be single valued if and only if the eigenvalues of the linear operator Δ:

$$(U, W) \longrightarrow ([\Omega_0, U] + [\Lambda, M_0] + [B, W]) , \ (W + [\Omega_0, W] + [\Lambda, P_0])$$

are integers. Indeed, if ξ denotes an eigenvalue of Δ, then the solutions of $\dfrac{dx}{dt} = \dfrac{1}{t}\Delta x(t)$ when restricted to the eigenspace corresponding to ξ are of the form $\dfrac{dx}{dt} = \dfrac{1}{t}\xi x(t)$, and hence, $x(t) = t^\xi x(1)$. But t^ξ is single valued only when ξ is a whole number.

To set the stage for the subsequent calculations observe that $\xi = -2$ is an eigenvalue of Δ. The demonstration is as follows: let $U = M_0$ and let $W = 2P_0$.

Since $[\Omega_0, M_0] + [B, P_0] = -M_0$, and $[\Omega_0, P_0] = -2P_0$

$$[\Omega_0, M_0] + [\Lambda, M_0] + [B, W] = 2[\Omega_0, M_0] + 2[B, P_0] = -2M_0$$

and

$$W + [\Omega_0, W] + [\Lambda, P_0] = 2P_0 + [\Omega_0, 2P_0] + [\Omega_0, P_0] = 2P_0 - 4P_0 - 2P_0 = -2(2P_0) .$$

Hence $\xi = -2$ is an eigenvalue.

The operator Δ is described by the following matrix

$$
\begin{pmatrix}
0 & \Delta_{32}r & \Delta_{32}q & 0 & b_3 & -b_2 \\[2mm]
\Delta_{13}r & 0 & \Delta_{13}p & -b_3 & 0 & b_1 \\[2mm]
\Delta_{21}q & \Delta_{21}p & 0 & b_2 & -b_1 & 0 \\[2mm]
0 & -\dfrac{h}{\lambda_2} & \dfrac{g}{\lambda_3} & 1 & \dfrac{r}{\lambda_3} & -\dfrac{q}{\lambda_2} \\[3mm]
\dfrac{h}{\lambda_1} & 0 & -\dfrac{f}{\lambda_3} & -\dfrac{r}{\lambda_3} & 1 & \dfrac{p}{\lambda_1} \\[3mm]
-\dfrac{g}{\lambda_1} & \dfrac{f}{\lambda_2} & 0 & \dfrac{q}{\lambda_2} & -\dfrac{p}{\lambda_1} & 1
\end{pmatrix}
$$

where $\Delta_{32} = \left(\dfrac{1}{\lambda_3} - \dfrac{1}{\lambda_1} \right)$, $\Delta_{13} = \left(\dfrac{1}{\lambda_1} - \dfrac{1}{\lambda_3} \right)$ and $\Delta_{21} = \left(\dfrac{1}{\lambda_2} - \dfrac{1}{\lambda_1} \right)$.

Consider now the spectrum of Δ associated with each solution in Lemma 4.

CASE 1. $p = r = 0$ $q = 2i\lambda_1$ $f = ih$, $g = 0$ $h = \dfrac{2\lambda_1}{b_3 + ib_1}$.

Here

$$
\Delta =
\begin{pmatrix}
0 & 0 & \Delta_{32}q & 0 & b_3 & 0 \\[2mm]
0 & 0 & 0 & -b_3 & 0 & b_1 \\[2mm]
0 & 0 & 0 & 0 & -b_1 & 0 \\[2mm]
0 & -\dfrac{h}{\lambda_2} & 0 & 1 & 0 & -\dfrac{q}{\lambda_2} \\[3mm]
\dfrac{h}{\lambda_1} & 0 & -\dfrac{f}{\lambda_3} & 0 & 1 & 0 \\[3mm]
0 & \dfrac{f}{\lambda_2} & 0 & \dfrac{q}{\lambda_2} & 0 & 1
\end{pmatrix} .
$$

Rename the variables $x_1 = u_2$, $x_2 = w_1$, $x_3 = w_3, x_4 = u_1, x_5 = u_3, x_6 = w_2$. Then,

$$
\dot{x}_1 = \dot{u}_2 = \frac{1}{t}(-b_3 x_2 + b_1 w_3),
$$
$$
\dot{x}_2 = \dot{w}_1 = \frac{1}{t}(-\frac{h}{\lambda_2}x_1 + x_2 - \frac{q}{\lambda_2}x_3),
$$
$$
\dot{x}_3 = \dot{w}_3 = \frac{1}{t}\left(\frac{f}{\lambda_2}x_1 + \frac{q}{\lambda_2}x_2 + x_3 \right) ,
$$

and

$$\dot{x}_4 = \dot{u}_1 = \frac{1}{t}(\Delta_{32} q x_5 + b_3 x_6),$$

$$\dot{x}_5 = \dot{u}_3 = \frac{1}{t}(-b_1 x_6),$$

$$\dot{x}_6 = \dot{w}_2 = \frac{1}{t}\left(\frac{h}{\lambda_1} x_4 - \frac{f}{\lambda_3} x_5 + x_6\right).$$

The preceding change of variables decomposes the variational equations into two independent subsystems and hence, the determinant of Δ is equal to the product of the determinants associated with

$$D_1 = \begin{pmatrix} 0 & -b_3 & b_1 \\ -\dfrac{h}{\lambda_2} & 1 & -\dfrac{q}{\lambda_2} \\ \dfrac{f}{\lambda_2} & \dfrac{q}{\lambda_2} & 1 \end{pmatrix}, \quad \text{and} \quad D_2 = \begin{pmatrix} 0 & \Delta_{32} q & b_3 \\ 0 & 0 & -b_1 \\ \dfrac{h}{\lambda_1} & -\dfrac{f}{\lambda_3} & 1 \end{pmatrix}.$$

It is easy to verify that the characteristic polynomial of D_1 is equal to $p_1(\xi) = -\xi(1-\xi)^2 + 6\xi - 6 = (\xi-1)(\xi-3)(\xi+2)$. The characteristic polynomial $p_2(\xi)$ associated with D_2 is given by

$$p_2(\xi) = \xi^2(1-\xi) + \xi\left(\frac{b_1 i h}{\lambda_3} + \frac{h b_3}{\lambda}\right) - \frac{2ihb_1}{\lambda_3} + \frac{2ihb_1}{\lambda}$$

where λ denotes the common eigenvalue $\lambda_1 = \lambda_2$.

Further, it is easy to check that $\xi = 2$ is a root of p_2, and therefore,

$$p_2(\xi) = (\xi-2)\left(-\xi(\xi+1) + \frac{1}{2}\left(\frac{2b_1 h}{\lambda_3} - \frac{2ihb_1}{\lambda}\right)\right).$$

$p_2(\xi)$ has integer roots only if

$$p(\xi) = -\xi(\xi+1) + \left(\frac{b_1 h i}{\lambda_3} - \frac{2hib_1}{\lambda}\right) = -\xi(\xi+1) + \frac{(\lambda-\lambda_3)}{\lambda\lambda_3}\frac{b_1(b_1+ib_3)\lambda}{b_1^2+b_3^2}$$

has integer roots.

Therefore, integer roots occur only when $b_3 = 0$ and $\dfrac{\lambda - \lambda_3}{\lambda_3}$ is an integer. The latter occurs only when the ratio $n = \dfrac{\lambda}{\lambda_3}$ is an integer.

We now pass to Case 2:

$$p = -iq, q(2\lambda_3 - \lambda_1) = 2\alpha_3 \lambda_1 \lambda_3, r = 2i\lambda_3, f = \frac{2\lambda_3}{b_1}, g = if, h = 0.$$

The case $2\lambda_3 - \lambda_1 = 0$ yields Kowalewski's case, and there is nothing further to prove. So assume that $2\lambda_3 - \lambda \neq 0$. Then $b_3 = 0$ implies that $q = 0$ and hence $p = 0$.

The corresponding differential system breaks up into two subsystem as in Case 1. Simply rename the variables $x_1 = u_1$ $x_2 = u_2$, $x_3 = w_3$, and $x_4 = u_3$, $x_5 = w_1$ $x_6 = w_2$.

The corresponding subsystems are described by the following matrices

$$D_1 = \begin{vmatrix} 0 & \Delta_{32}r & 0 \\ \Delta_{13}r & 0 & b_1 \\ -\dfrac{if}{\lambda} & \dfrac{f}{\lambda} & 1 \end{vmatrix}, \quad \text{and} \quad D_2 = \begin{vmatrix} 0 & 0 & -b_1 \\ \dfrac{if}{\lambda_3} & 1 & \dfrac{r}{\lambda_3} \\ -\dfrac{f}{\lambda_3} & -\dfrac{r}{\lambda_3} & 1 \end{vmatrix}.$$

Then the characteristic polynomial $p_1(\xi)$ of D_1 is given by

$$p_1(\xi) = \xi^2(1-\xi) + \xi\left(\Delta_{32}\Delta_{13}r^2 + \frac{b_1 f}{\lambda}\right) - \Delta_{32}\Delta_{13}r^2 - \frac{\Delta_{32}rifb_1}{\lambda}.$$

It follows that

$$\Delta_{32}\Delta_{13}r^2 = \frac{4(\lambda - \lambda_3)^2}{\lambda^2}, \quad \text{and that} \quad \frac{b_1 f}{\lambda} = \frac{b_1\lambda_3}{\lambda^2}.$$

Furthermore,

$$\frac{\Delta_{32}rifb_1}{\lambda} = -\frac{4(\lambda - \lambda_3)\lambda_3}{\lambda_2}.$$

Therefore,

$$p_1(\xi) = \xi^2(1-\xi) + \frac{\xi}{\lambda^2}(4(\lambda - \lambda_3)^2 + \lambda_1\lambda_3) - \frac{4(\lambda - \lambda_3)(\lambda - 2\lambda_3)}{\lambda_2}.$$

For $p_1(\xi)$ to have integer roots the quantity $\dfrac{4(\lambda - \lambda_3)(\lambda - 2\lambda_3)}{\lambda_2}$ must be an integer.

We have already shown that $n = \dfrac{\lambda_3}{\lambda}$ must be an integer. Therefore, $4\left(1 - \dfrac{1}{n}\right)\left(1 - \dfrac{2}{n}\right)$ must be an integer. This quantity is an integer only when $n = 1$, or $n = 2$. Then $n = 1$ implies that $\lambda_1 = \lambda_2 = \lambda_3$, and $n = 2$ implies that $\lambda_1 = \lambda_2 = 2\lambda_3$ and $b_3 = 0$. This ends the proof of Theorem 3 for $\epsilon = 0$. The rest of the proof follows by an argument identical to the one used in the previous theorem.

The remainder of the paper shows that each of the three possibilities for meromorphic solutions produces an extra integral of motion, exactly as in the case of mechanical tops. Remarkably, it turns out that the resulting equations can be solved by quadrature and that the solutions are meromorphic functions of complex time.

However, this fact does not necessarily rule out the possibility of existence of other integrals of motion outside of those singled out by the Kowalewski-Lyapunov criteria ([**Lvl**]).

VIII. Cartan algebras, root spaces and extra integrals of motion

We now continue with our investigations of the meromorphic solutions of the Hamiltonian equations associated with

$$H = \frac{1}{2}\left(\frac{M_1^2}{\lambda_1} + \frac{M_2^2}{\lambda_2} + \frac{M_3^2}{\lambda_3}\right) + b_1 p_1 + b_2 p_2 + b_3 p_3 \ .$$

under the assumption that at least two eigenvalues λ_1, λ_2, λ_3 are equal. Each equality among the eigenvalues determines a particular Cartan subalgebra of $so_4(\mathbb{C})$ that illuminates the geometry of the solutions. For concreteness we shall assume that $\lambda_1 = \lambda_2$, although each other choice is equally valid. This choice of equal eigenvalues singles out the commutative algebra spanned by the basis elements B_3 and A_3 in Table 1.

Recall that a Cartan algebra $\mathfrak{h}$ is a maximal commutative sub algebra of a Lie algebra $\mathfrak{g}$. It is known that all Cartan sub-algebras of a semi-simple Lie algebra are conjugate, and hence all have the same dimension. The dimension of any Cartan algebra is called the rank of $\mathfrak{g}$. The rank of $so_4(\mathbb{C})$ is two.

An element α in the dual $\mathfrak{h}^*$ of a Cartan algebra is called a root if for some $v \in \mathfrak{g}$, $[x, v] = \alpha(x)v$ for all $x \in \mathfrak{h}$. An easy calculation, which we shall omit, indicates that there are four distinct roots α_1, $-\alpha_1$, α_2, $-\alpha_2$, defined as follows

$$\alpha_1(xA_3 + yB_3) = i(x + y) \ , \quad \text{and} \quad \alpha_2(xA_3 + yB_3) = i(x - y) \ .$$

The corresponding root spaces are one dimensional and are generated by

$$v_1 = \frac{1}{2}(A_1 + iA_2) + \frac{1}{2}(B_1 + iB_2) \qquad v_3 = \frac{1}{2}(A_1 + iA_2) - \frac{1}{2}(B_1 + iB_2)$$

$$v_2 = \frac{1}{2}(A_1 - iA_2) + \frac{1}{2}(B_1 - iB_2) \qquad v_4 = \frac{1}{2}(A_1 - iA_2) - \frac{1}{2}(B_1 - iB_2)$$

To these four matrices we adjoin

$$v_5 = -\frac{i}{2}(A_3 + B_3) \ , \quad \text{and} \quad v_6 = -\frac{i}{2}(A_3 - B_3) \ .$$

It follows that $\alpha_1(v_5) = 1$ and $\alpha_2(v_6) = 1$. Consequently,

$$[v_5, v_1] = \alpha_1(v_5)v_1 = v_1, [v_5, v_2] = -\alpha_1(v_5)v_2 = -v_2,$$

$$[v_6, v_3] = \alpha_2(v_6)v_3 = v_3.[v_6, v_4] = -\alpha_2(v_6)v_4 = -v_4.$$

Furthermore,

$$[v_1, v_2] = i(A_3 + B_3) = -2v_5 , \quad \text{and} \quad [v_3, v_4] = i(A_3 - B_3) = -2v_6 .$$

Thus, each of $\{v_1, v_2, v_5\}$ and $\{v_3, v_4, v_6\}$ form a basis for a 3-dimensional Lie sub algebra of $sl_4(\mathbb{C})$. Each of these subalgebras is isomorphic to $sl_2(\mathbb{C})$ as can be most directly seen from the isomorphism $\phi = \Phi_*$ established in Proposition 6.1 (ii). Relations (47) and (48) imply that

$$\phi(v_1 + v_2) = (E_1, 0), \ \phi(v_1 - v_2) = (iE_2, 0), \ \phi(v_5) = -(2iE_3, 0)$$
$$\phi(v_3 + v_4) = (0, E_1), \ \phi(v_3 - v_4) = (0, iE_2), \ \phi(v_6) = (0, -2iE_3)$$

We will now express the Hamiltonian equations in terms of the dual coordinates relative to the basis v_1, v_2, v_5, v_3, v_4, v_6.

Let $x_1v_1^* + x_2v_2^* + x_3v_5^* + y_1v_3^* + y_2v_4^* + y_3v_6^*$ denote a general point in $so_4^*(\mathbb{C})$.

Since

$$A_1 = \frac{1}{2}(v_1 + v_2) + \frac{1}{2}(v_3 + v_4) \quad , \quad B_1 = \frac{1}{2}(v_1 + v_2) - \frac{1}{2}(v_3 + v_4)$$

$$A_2 = \frac{1}{2i}(v_1 - v_2) + \frac{1}{2i}(v_3 - v_4) \quad , \quad B_2 = \frac{1}{2i}(v_1 - v_2) - \frac{1}{2i}(v_3 - v_4)$$

$$A_3 = i(v_5 + v_6) \quad , \quad B_3 = i(v_5 - v_6) ,$$

it follows that

$$M_1 = \frac{1}{2}(x_1 + x_2) + \frac{1}{2}(y_1 + y_2) \quad , \quad p_1 = \frac{1}{2}(x_1 + x_2) - \frac{1}{2}(y_1 + y_2)$$

$$M_2 = \frac{1}{2i}(x_1 - x_2) + \frac{1}{2i}(y_1 - y_2) \quad , \quad p_2 = \frac{1}{2i}(x_1 - x_2) - \frac{1}{2i}(y_1 - y_2)$$

$$M_3 = i(x_3 + y_3) \quad , \quad p_3 = i(x_3 - y_3) .$$

Therefore,

$$M_1 + iM_2 = x_1 + y_1 \qquad p_1 + ip_2 = x_1 - y_1$$
$$M_1 - iM_2 = x_2 + y_2 \qquad p_1 - ip_2 = x_2 - y_2 .$$

The Hamiltonian $\mathcal{H} = \frac{1}{2\lambda}(M_1^2 + M_2^2) + \frac{M_3^2}{2\lambda_3} + b_1p_1 + b_2p_2 + b_3p_3$ takes the following form in the new coordinates:

$$\mathcal{H} = \frac{1}{2\lambda}(x_1 - y_1)(x_2 - y_2) - \frac{1}{2}\frac{(x_3 + y_3)^2}{\lambda_3} + \frac{1}{2}b(x_2 + y_2) + \frac{1}{2}\bar{a}(x_1 + y_1) + ib_3(x_3 - y_3)$$

where $b = b_1 + ib_2$, and $\bar{b} = b_1 - ib_2$.

The Poisson algebra spanned by x_1, x_2, x_3 is isomorphic to the algebra spanned by v_1, v_2, v_5, and an analogous statement applies to the Poisson algebra spanned by y_1, y_2, y_3. Therefore,

$$\{x_1, x_2\} = -2x_3 \ , \ \{x_3, x_1\} = x_1 \ , \ \{x_3, x_2\} = -x_2 \ , \quad \text{and}$$

$$\{y_1, y_2\} = -2y_3 \ , \ \{y_3, y_1\} = y_1 \ , \ \{y_3, y_2\} = -y_2 \ , \quad \text{otherwise}$$

$$\{x_i, y_j\} = 0 \quad \text{for all} \quad i \text{ and } j \ .$$

Then,

$$\frac{dx_1}{dt} = \{x_1, \mathcal{H}\} = -\frac{(x_1 + y_1)x_3}{\lambda} + \frac{(x_3 + y_3)}{\lambda_3}x_1 - bx_3 - ib_3x_1$$

$$\frac{dx_2}{dt} = \{x_2, \mathcal{H}\} = \frac{x_3}{\lambda}(x_2 + y_2) - \frac{(x_3 + y_3)x_2}{\lambda_3} + \bar{b}x_3 + ib_3x_2$$

(63)

$$\frac{dx_3}{dt} = \{x_3, \mathcal{H}\} = -\frac{1}{2\lambda}(x_1 + y_1)x_2 + \frac{1}{2\lambda}(x_2 + y_2)x_1 - \frac{1}{2}bx_2 + \frac{1}{2}\bar{b}x_1 \ .$$

Similarly,

$$\frac{dy_1}{dt} = \{y_1, \mathcal{H}\} = -\frac{(x_1 + y_1)}{\lambda}y_3 + \frac{(x_3 + y_3)}{\lambda_3}y_1 + by_3 + ib_3y_1$$

$$\frac{dy_2}{dt} = \{y_2, \mathcal{H}\} = \frac{(x_2 + y_2)}{\lambda}y_3 - \frac{(x_3 + y_3)}{\lambda_3}y_2 - \bar{b}y_3 - ib_3y_2$$

(64)

$$\frac{dy_3}{dt} = \{y_3, \mathcal{H}\} = -\frac{1}{2\lambda}(x_1 + y_1)y_2 + \frac{1}{2\lambda}(x_2 + y_2)y_1 + \frac{1}{2}by_2 - \frac{1}{2}\bar{b}y_1 \ .$$

We now address the existence of additional integrals of motion for the meromorphic cases classified in the previous section.

Let us consider first the cases of Lagrange:

CASE 1. $\lambda_1 = \lambda_2 = \lambda_3$, A_0 arbitrary, and

CASE 2. $\lambda_1 = \lambda_2$, λ_3 arbitrary, but $\alpha_1 = \alpha_2 = 0$.

In these situations the original equations $\frac{d\hat{M}}{dt} = \hat{M} \times \hat{\Omega} + \hat{P} \times \hat{B}$ and $\frac{d\hat{P}}{dt} = \hat{P} \times \hat{\Omega} + \epsilon\hat{M} \times \hat{B}$ immediately yield two integrals of motion: $I_4 = \hat{M} \cdot \hat{A}_0 = M_1\alpha_1 + M_2\alpha_2 + M_3\alpha_3$ for Case 1, and $I_4 = M_3$ for Case 2. By rotating the constants b_1, b_2, b_3 into $b_1 = b_2 = 0$ $b_3 \neq 0$, Case 1 can be brought to Case 2, so we shall simply refer to $\lambda_1 = \lambda_2$, $b_1 = b_2 = 0$ as the case of Lagrange. The existence of this integral of motion can also be deduced from equations (63) and (64) as follows:

When $b_1 = b_2 = 0$

$$\frac{d}{dt}(x_3 + y_3) = \frac{1}{2\lambda}\big((x_2 + y_2)(x_1 + y_1) - (x_2 + y_2)(x_1 + y_1)\big) = 0$$

and so $x_3 + y_3 =$ constant. But $x_3 + y_3 = -iM_3$, and therefore, $M_3(t) =$ constant.

Let us now turn to the case of Kowalewski: $\lambda = \lambda_1 = \lambda_2 = 2\lambda_3$ and $b_3 = 0$.

It will be convenient to introduce the following notations

$$z_1 = \frac{1}{2}(x_1 + y_1) = \frac{1}{2}(M_1 + iM_2) \qquad z_2 = \frac{1}{2}(x_2 + y_2) = \frac{1}{2}(M_1 - iM_2)$$
$$w_1 = x_1 - y_1 = p_1 + ip_2 \qquad w_2 = x_2 - y_2 = p_1 - ip_2 \qquad .$$

Since

$$\frac{dx_1}{dt} = -\frac{2z_1}{\lambda}x_3 + \frac{(x_3 + y_3)}{\lambda_3}x_1 - bx_3$$

$$\frac{dy_1}{dt} = \frac{2z_1}{\lambda}y_3 + \frac{(x_3 + y_3)}{\lambda_3}y_1 - by_3$$

$$\frac{dx_2}{dt} = \frac{2z_2}{\lambda}x_3 - \frac{(x_3 + y_3)}{\lambda_3}x_2 + \bar{b}x_3$$

$$\frac{dy_2}{dt} = -\frac{2z_2}{\lambda}y_3 - \frac{(x_3 + y_3)}{\lambda_3}y_2 + \bar{b}y_3$$

$$2\frac{dz_1}{dt} = -\frac{2z_1}{\lambda}(x_3 + y_3) + 2\frac{(x_3 + y_3)z_1}{\lambda_3} - b(x_3 - y_3).$$

After the substitutions $\lambda = 2\lambda_3$, $x_3 + y_3 = -iM_3$, $x_3 - y_3 = -ip_3$ into the above equations we get

$$2\frac{dz_1}{dt} = -\frac{iM_3}{\lambda_3}z_1 + ibp_3.$$

Similarly,

$$2\frac{dz_2}{dt} = \frac{iM_3}{\lambda_3}z_2 - i\bar{b}p_3,$$

and

$$\frac{dw_1}{dt} = \frac{z_1 ip_3}{\lambda_3} - \frac{iM_3 w_1}{\lambda_3} + ibM_3,$$

$$\frac{dw_2}{dt} = \frac{z_2 ip_3}{\lambda_3} - \frac{iM_3}{\lambda_3}w_2 - i\bar{b}M_3 .$$

Therefore,

$$\frac{d}{dt}\left(\frac{1}{\lambda_3}z_1^2\right) = -\frac{iM_3 z_1^2}{\lambda_3^2} + \frac{ibh_3 z_1}{\lambda_3}$$

$$\frac{d}{dt}bw_1 = \frac{bz_1 ih_3}{\lambda_3} - \frac{ibM_3 w_1}{\lambda_3} + ib^2 M_3,$$

and consequently,

$$\frac{d}{dt}\left(\frac{1}{\lambda_3}z_1^2 - bw_1 + \lambda_3 b^2\right) = -\frac{iM_3 z_1^2}{\lambda_3^2} + \frac{ibM_3 w_1}{\lambda_3} - ib^2 M_3$$

$$= -\frac{iM_3}{\lambda_3}\left(\frac{z_1^2}{\lambda_3} - bw_1 + \lambda_3 b^2\right).$$

Similarly,

$$\frac{d}{dt}\frac{z_2^2}{\lambda_3} = \frac{iM_3 z_1^2}{\lambda_3^2} + \frac{i\bar{b}p_3 z_2}{\lambda_3}, \quad \text{and}$$

$$\frac{d}{dt}\bar{b}w_2 = \frac{ip_3 z_2 \bar{b}}{\lambda_3} + \frac{i\bar{b}M_3 w_2}{\lambda_2} - i\bar{b}^2 M_3,$$

and

$$\frac{d}{dt}\left(\frac{z_2^2}{\lambda_3} - \bar{b}w_2 + \bar{b}\lambda_3\right) = \frac{iM_3 z_2^2}{\lambda_3^2} - \frac{i\bar{b}M_3 w_2}{\lambda_3} + i\bar{b}^2 M_3 = \frac{iM_3}{\lambda_3}\left(\frac{z_2^2}{\lambda_3} - \bar{b}w_2 + \bar{b}^2\lambda_3\right).$$

Therefore,

$$I = \left(\frac{z_1^2}{\lambda_3} - bw_1 + b^2\lambda_3\right)\left(\frac{z_2^2}{\lambda_3} - \bar{b}w_2 + \bar{b}^2\lambda_3\right)$$

is an integral of motion for $\mathcal{H}$.

The function I is a constant multiple of

(65)
$$I_4 = \left(z_1^2 - a(w_1 - a)\right)\left(z_2^2 - \bar{a}(w_2 - \bar{a})\right)$$

with $a = \lambda b$.

This integral of motion takes on different forms when restricted to the real forms of $so_4(\mathbb{C})$:

1. on $so_4(\mathbb{R})$,
$$I_4 = |z_1^2 - a(w_1 - a)|^2$$
because the variables $M_1, M_2, M_3, p_1, p_2, p_3$ are all real and $z_2 = \bar{z}_1$ and $w_2 = \bar{w}_1$.

2. on $so(1,3)$,
$$I_4 = |z_1^2 - a(w_1 + a)|^2.$$
In this case the variables M_1, M_2, M_3 are real, while p_1, p_2, p_3 are imaginary and need to be replaced by ip_1, ip_2, ip_3 (see Table 2). For the same reason a need to be replaced by $-ia$. Then $z_2 = \bar{z}_1$ and $w_1 = i(p_1 + ip_2)$ and $w_2 = i(p_1 - ip_2)$. The above form results when w_1 and w_2 are replaced by iw_1 and $i\bar{w}_1$.

3. It is easy to verify that $I_4 = (z_1^2 - aw_1)(z_2^2 - \bar{a}w_2)$ is an integral of motion for the semi-direct product. This fact can be either verified directly, or it

can be verified using the deformation procedure:

$$I_4(\epsilon) = \left(z_1^2 - \epsilon a \left(\frac{1}{\epsilon}w_1 - \epsilon a\right)\right) \left(z_2^2 - \epsilon \bar{a} \left(\frac{1}{\epsilon}w_2 - \epsilon a\right)\right)$$

is an integral of motion for the deformed Lie algebra $\mathfrak{g}_\varepsilon$. The integral of motion on the semi-direct product is the limit as ϵ tends to zero. As in the above cases $I_4 = |z_1^2 - aw_1|$ when M_1, M_2, M_3 and p_1, p_2, p_3 are restricted to real values.

4. In all these cases I_4 can be described by a single parameter ϵ as

$$I_4 = |z^2 - a(w - \epsilon a)|$$

that is an integral of motion for the Hamiltonian

$$H = \frac{1}{\lambda}(M_1^2 + M_2^2) + \frac{1}{\lambda_3}M_3^2 + b_1 p_1 + b_2 p_2$$

([**Ju3**] and ([**KoK**])).

5. The remaining case, not usually mentioned in the literature on the top, is the case of $so(2, 2)$. Here

$$H = \frac{1}{\lambda}(M_1^2 - M_2^2) - \frac{1}{\lambda_3}M_3^2 + b_1 p_1 + b_2 p_2$$

and

$$I_4 = (\frac{1}{4}(M_1 - M_2)^2 - a(p_1 - p_2 - a))(\frac{1}{4}(M_1 + M_2)^2 - a(p_1 + p_2 - a))$$

where $a = \lambda(b_1 + b_2)$.

Hence I_4 of this paper may be regarded as a holomorphic extension of the integral of motion presented in [**Ju3**] and [**KoK**].

It follows from the above that each meromorphic case is also completely integrable in the sense of Louiville, for in addition to the Hamiltonian $\mathcal{H} = I_1$, there are two Casimir functions

$$I_2 = \epsilon(M_1^2 + M_2^2 + M_3^2) + p_1^2 + p_2^2 + p_3^2 \ , \ I_3 = p_1 M_1 + p_2 M_2 + p_3 M_3$$

which together with I_4 form an involutive family of functions on the Lie algebra of G. Since the Hamiltonians of right-invariant vector fields commute with the Hamiltonians of left-invariant vector fields on G, it follows that these Hamiltonians, which are functions on the entire product $\mathfrak{g}^* \times G$, Poisson commute with any left-invariant function I. The maximal number of commuting right-invariant Hamiltonians is equal to the rank of $\mathfrak{g}$ which is two on $SO_4(\mathbb{C})$. Therefore, $\mathcal{H}$ is completely integrable whenever there are four independent functions on $\mathfrak{g}^*$ in involution with each other.

It remains to show that each of these integrable cases can be explicitly integrated by quadrature and that the corresponding solutions are meromorphic solutions of complex time.

IX. Elastic Curves for the case of Lagrange

To maintain continuity of notations with our previous exposition on this topic ([**Ju2**]) it will be assumed that $\lambda_2 = \lambda_3$ (instead of $\lambda_1 = \lambda_2$). It will be also assumed that $b_1 = 1$, $b_2 = b_3 = 0$. Then the Hamiltonian is given by

$$\mathcal{H} = \frac{1}{2}\frac{M_1^2}{\lambda_1} + \frac{1}{2\lambda}(M_2^2 + M_3^2) + p_1$$

with λ denoting the common value $\lambda_2 = \lambda_3$.

We shall continue to work with the coordinates x_1, x_2, x_3, y_1, y_2, y_3 introduced in the previous section. Because of the permutation of indexes these coordinates are now defined as follows:

$$M_1 = i(x_1 + y_1) \qquad , \qquad p_1 = i(x_1 - y_1)$$

$$M_2 = \frac{1}{2}(x_2 + x_3) + \frac{1}{2}(y_2 + y_3) \qquad , \qquad p_2 = \frac{1}{2}(x_2 + x_3) - \frac{1}{2}(y_2 + y_3)$$

$$M_3 = \frac{1}{2i}(x_2 - x_3) + \frac{1}{2i}(y_2 - y_3) \qquad , \qquad p_3 = \frac{1}{2i}(x_2 - x_3) - \frac{1}{2i}(y_2 - y_3) \ .$$

The Cartan algebra is given by the span of A_1 and B_1, hence the new coordinates conform to the following Poisson brackets

$$\{x_1, x_2\} = x_2 \ , \ \{x_1, x_3\} = -x_3 \ , \ \{x_2, x_3\} = -2x_1 \ , \quad \text{and}$$

$$\{y_1, y_2\} = y_2 \ , \ \{y_1, y_3\} = -y_2 \ , \ \{y_2, y_3\} = -2y_1 \ , \quad \text{and}$$

$$\{x_i, y_j\} = 0 \quad \text{for all } i \text{ and } j.$$

It is advantageous to work with the coordinates z_1, z_2, w_1, w_2, p_1, M_1 defined by:

$$z_1 = M_2 + iM_3 = x_2 + y_2, \ z_2 = M_2 - iM_3 = x_3 + y_3$$

$$w_1 = p_2 + ip_3 = x_2 - y_2, \ w_2 = p_2 - ip_2 = x_3 - y_3.$$

The following Poisson bracket table will be handy

$\{\ ,\ \}$	p_1	M_1	z_1	z_2	w_1	w_2
p_1	0	0	iw_1	$-iw_2$	$i\epsilon z_1$	$-i\epsilon z_2$
M_1	0	0	iz_1	$-iz_2$	iw_1	$-iw_2$
z_1	$-iw_1$	$-iz_1$	0	$2iM_1$	0	$2ip_1$
z_2	iw_2	iz_2	$-2iM_1$	0	$-2ip_1$	0
w_1	$-i\epsilon z_1$	$-iw_1$	0	$2ip_1$	0	$2i\epsilon M_1$
w_2	$i\epsilon z_2$	iw_2	$-2ip_1$	0	$-2i\epsilon M_1$	0

TABLE 5

where

$\epsilon = 1$ corresponds to the $so_4(\mathbb{C})$, while $\epsilon = 0$ corresponds to the semi-direct product.

The above table immediately shows that the Hamiltonian equations associated with $\mathcal{H}$ are given by:

$$\frac{dz_1}{dt} = -i\left(\frac{1}{\lambda_1} - \frac{1}{\lambda}\right) M_1 z_1 - iw_1$$

$$\frac{dz_2}{dt} = i\left(\frac{1}{\lambda_1} - \frac{1}{\lambda}\right) M_1 z_2 + iw_2$$

$$\frac{dw_1}{dt} = -\frac{iM_1 w_1}{\lambda_1} + iz_1 \left(\frac{1}{\lambda}p_1 - \epsilon\right)$$

(66)
$$\frac{dw_2}{dt} = \frac{iM_1 w_2}{\lambda_1} - iz_2 \left(\frac{1}{\lambda}p_1 - \epsilon\right) .$$

It follows from the above that

$$\frac{d}{dt} z_1 z_2 = i(w_2 z_1 - w_1 z_2)$$

(67)
$$\frac{d}{dt}(w_1 w_2) = i\left(\frac{p_1}{\lambda} - \epsilon\right)(w_2 z_1 - w_1 z_2).$$

Let $u = w_2 z_1 - w_1 z_2$, and let $x = z_1 z_2$. We shall presently see that u and x can be identified with points on an elliptic curve, an identification that plays central importance for the integration of the basic equations. The coefficients of this elliptic curve are defined by integrals of motion:

1. The reduced Hamiltonian $I_1 = \mathcal{H} - \dfrac{M_1^2}{2\lambda_1} = \dfrac{1}{2\lambda}z_1 z_2 + p_1 = \dfrac{1}{2\lambda}x + p_1.$

2. The Casimir integral $I_2 = p_1 M_1 + p_2 M_2 + p_3 M_3 = p_1 M_1 + \dfrac{1}{2}z_2 w_1 + \dfrac{1}{2}z_1 w_2.$

3. Another (reduced) Casimir integral $I_3 = p_1^2 + p_2^2 + p_3^2 + \epsilon(M_1^2 + M_2^2 + M_3^2) - \epsilon M_1^2 = p_1^2 + \epsilon z_1 z_2 + w_1 w_2 .$

4. The conserved quantity $I_4 = M_1.$

Then, along each extremal curve the variables $u(t)$ and $x(t)$ evolve as follows:

$$u^2(t) = (w_1(t)z_2(t) - w_2(t)z_1(t))^2 = (w_1(t)z_2(t) + w_2(t)z_1(t))^2 - 4z_1(t)z_2(t)w_1(t)w_2(t)$$
$$= 4(I_2 - p_1(t)M_1)^2 - 4x(t)\left(I_3 - p_1^2(t) - \epsilon x(t)\right)$$
$$= 4(I_2 - (I_1 - \frac{1}{2\lambda}x(t))M_1)^2 - 4x(t)\left(I_3 - (I_1 - \frac{1}{2\lambda}x(t))^2 - \epsilon x(t)\right)$$

Hence,

$$u^2(t) = \frac{1}{\lambda^2}x^3(t) + Ax^2(t) + Bx(t) + C$$

where $A = \dfrac{M_1^2}{\lambda^2} - 4\left(\dfrac{I_1}{\lambda} - \epsilon\right)$, $B = 4\left(\dfrac{M_1}{\lambda}(I_2 - I_1 M_1) - (I_3 - I_1^2)\right)$, and where $C = 4(I_2 - M_1 I_1)^2$.

We shall use Γ to denote the elliptic curve

(68)
$$u^2 = \frac{1}{\lambda^2}x^3 + Ax^2 + Bx + C .$$

The curve Γ is equivalent to the cubic curve of Weierstrass

$$v^2 = \frac{1}{4}\xi^3 - g_2\xi - g_3$$

with $v = \dfrac{1}{2}\lambda u$ and $\xi = x + \dfrac{1}{3}A\lambda$.

It follows from (67) that $\dfrac{dx}{dt}(t) = iu(t)$ and therefore,

$$\left(\frac{dx}{dt}\right)^2 = -u^2(t) = -\left(\frac{1}{\lambda^2}x^3(t) + Ax^2(t) + Bx(t) + C\right) .$$

The above implies that the roots of the polynomial $P(x) = \dfrac{1}{\lambda^2}x^3 + Ax^2 + Bx + C$ correspond to the stationary points of $x(t)$. This observation implies that the quantity $z_1(t)z_2(t)$ is constant whenever it originates on the surface $z_1 w_2 - z_2 w_1 = 0$.

But if $z_1(t)z_2(t)$ is constant, then $p_1(t) = I_1 - \dfrac{z_1 z_2}{2\lambda}$ is constant, and $w_1 w_2 = I_3 - p_1^2 - \epsilon z_1 z_2$ is also constant. The reader may note that these stationary solutions are confined to the singular points of the algebraic variety $V(I_1, I_2, I_3, I_4)$ defined by $I_1 = $ constant, $I_2 = $ constant, $I_3 = $ constant, and $I_4 = $ constant. We leave this verification to the reader.

We are now ready to outline the integration procedure. As we have already shown

$$\left(\frac{dx}{dt}\right)^2 = -u^2(t) = -\left(\frac{1}{\lambda^2}x^3(t) + Ax^2 + Bx + C\right) .$$

Apart from the equilibrium points described earlier, each solution $x(t)$ is a non-constant elliptic function, hence meromorphic.

In fact every solution of

$$\left(\frac{dx}{dt}\right)^2 = -u^2 = -\left(\frac{1}{\lambda^2}x^3 + Ax^2 + Bx + C\right)$$

can be expressed in terms of the Weierstrass $\wp$-function.

Simply define $\xi(t) = x(\frac{\lambda}{2}it) + \dfrac{A\lambda^2}{3}$. Then

$$\begin{aligned}
\left(\frac{d\xi}{dt}\right)^2 &= -\frac{1}{4}\lambda^2\left(\frac{dx}{dt}\right)^2 \\
&= \frac{1}{4}x^3 + 4\lambda^2 Ax^2 + 4\lambda^2 B + 4\lambda^2 C \\
&= \frac{1}{4}\xi^3(t) - g_2\xi(t) - g_3 \ .
\end{aligned}$$

Hence, $\xi(t)$ is a time-shift of $\wp$. For simplicity of exposition we shall assume that $\xi(t)$ is time shifted so that the pole is at $t = 0$. Then $x(t)$ is an even elliptic function having a double pole at $t = 0$. Since $p_1(t) = I_1 - \dfrac{1}{2\lambda}x(t)$, $p_1(t)$ is also an even elliptic function having a double pole at $t = 0$.

The remaining variables are integrated as follows. Since $I_2 = p_1 M_1 + \dfrac{1}{2}(z_1 w_2 + z_2 w_1)$ and $\dfrac{1}{2}u = \dfrac{1}{2}(z_1 w_2 - z_2 w_1)$ it follows that $z_1 w_2 = (I_2 - p_1 M_1) + \dfrac{1}{2}u$, and $z_1 w_1 = (I_2 - p_1 M_1) - \dfrac{1}{2}u$.

Assuming that $w_1(t)w_2(t) \neq 0$ then

$$\frac{z_1}{w_1} = \frac{z_1 w_2}{w_1 w_2} = \frac{(I_2 - p_1 M_1) + \dfrac{1}{2}u}{I_3 - p_1^2 - \epsilon x} \quad \text{and}$$

$$\frac{z_2}{w_2} = \frac{z_2 w_1}{w_2 w_1} = \frac{(I_2 - p_1 M_1) - \dfrac{1}{2}u}{I_3 - p_1^2 - \epsilon x} \ .$$

and therefore,

$$z_1(t) = w_1(t)\frac{(I_2 - p_1(t)M_1) + \dfrac{1}{2}u(t)}{I_3 - p_1^2(t) - \epsilon x(t)}$$

and

$$z_2(t) = w_2(t)\frac{(I_2 - p_1(t)M_1) - \dfrac{1}{2}u(t)}{I_3 - p_1^2(t) - \epsilon x(t)} \ .$$

Recall (equations (66)) that

$$\frac{dw_1}{dt} = -i\left(\frac{M_1 w_1}{\lambda_1} - z_1\left(\frac{1}{\lambda}p_1 - \epsilon\right)\right) \ .$$

Hence,

$$(69) \quad \frac{dw_1}{dt} = -iw_1 \left(\frac{M_1}{\lambda_1} - \frac{\left(\left(I_2 - p_1 M_1\right) + \frac{1}{2}u \right)}{\left(I_3 - p_1^2 - \epsilon x\right)} \left(\frac{p_1}{\lambda} - \epsilon \right) \right) = -iw_1(t) f(t)$$

where

$$f(t) = \frac{M_1}{\lambda_1} - \frac{\left(\left(I_2 - p_1 M_1\right) + \frac{1}{2}u \right)}{\left(I_3 - p_1^2 - \epsilon x\right)} \left(\frac{p_1}{\lambda} - \epsilon \right) \ .$$

It follows that $w_1(t) = (\exp -i \int_0^t f(z)dz) w_1(0)$, provided that $\int_0^t f(z)dz$ is independent of the path that connects 0 to t; that is, provided that f has zero residue. Since u is an odd function, it is not at all obvious that the path integral of f is single valued . To show that $w_1(t)$ is a meromorphic function we need to proceed differently .

Since $I_1 = \frac{x}{2\lambda} + p_1$, and $I_3 = p_1^2 + \epsilon x + w_1 w_2$ it follows that $p_1^2 + (I_1 - p_1)2\lambda\epsilon + w_1 w_2 = I_3$, or that $p_1^2 - 2\epsilon\lambda p_1 + w_1 w_2 = I_3 - 2\epsilon\lambda I_1$. Hence, $(p_1 - \lambda\epsilon)^2 + w_1 w_2 = M^2$ where M is another constant of motion defined by $M^2 = I_3 - 2\epsilon\lambda I_1 + \lambda^2\epsilon^2$.

Now introduce an angle θ through the formula $p_1 - \lambda\epsilon = M\cos\theta$. Since

$$\frac{dp_1}{dt} = -\frac{1}{2\lambda}\frac{dx}{dt} == -\frac{1}{2\lambda}iu \ ,$$

it follows that

$$M\sin\theta\frac{d\theta}{dt} = \frac{i}{2\lambda}u(t).$$

Note that $w_1 w_2 = M^2 - (p_1 - \epsilon\lambda)^2 = M^2\sin^2\theta$. Then,

$$\begin{aligned}
\frac{M_1}{\lambda_1} - f(t) &= \frac{\left(I_2 - p_1 M_1 + \frac{1}{2}u\right)}{w_1 w_2}\left(\frac{p_1}{\lambda} - \epsilon\right) = \frac{\left(I_2 - p_1 M_1 + \frac{1}{2}u\right)M\cos\theta}{\lambda M^2\sin^2\theta} \\
&= \frac{(I_2 - p_1 M_1)M\cos\theta}{\lambda M^2\sin^2\theta} + \left(\frac{\lambda}{i}M\sin\theta\frac{d\theta}{dt}\right)\frac{M\cos\theta}{\lambda M^2\sin^2\theta} \\
&= \frac{(I_2 - p_1 M_1)\cos\theta}{\lambda M\sin^2\theta} - i\frac{\cos\theta}{\sin\theta}\left(\frac{d\theta}{dt}\right) \ .
\end{aligned}$$

Now introduce another complex function $\varphi(t)$ through the formula

$$w_1(t) = M\sin\theta(t)e^{-i\left(\varphi(t) + t\frac{M_1}{\lambda_1}\right)} ,$$

Then,

$$-iw_1(t)f(t) = \frac{dw_1}{dt} = \left(M\cos\theta\frac{d\theta}{dt} - i\left(\frac{d\varphi}{dt} + \frac{M_1}{\lambda_1}\right)M\sin\theta\right)e^{i\left(\varphi + \frac{M_1(t)}{\lambda_1}\right)}$$

$$= \left(M\cos\theta\frac{d\theta}{dt} - i\left(\frac{d\varphi}{dt} + \frac{M_1}{\lambda_1}\right)M\sin\theta\right)\frac{w_1(t)}{M\sin\theta}$$

Hence,

$$-if(t)M\sin\theta = M\cos\theta\frac{d\theta}{dt} - i\left(\frac{d\varphi}{dt} + \frac{M_1}{\lambda_1}\right)M\sin\theta\ ,$$

or

$$-i\left(f(t) - \frac{M_1}{\lambda_1}\right)M\sin\theta = M\cos\theta\frac{d\theta}{dt} - i\frac{d\varphi}{dt}M\sin\theta.$$

After the substitution of $\dfrac{M_1}{\lambda_1} - f(t) = \dfrac{(I_2 - p_1 M_1)\cos\theta}{\lambda M \sin^2\theta} - i\dfrac{\cos\theta}{\sin\theta}\left(\dfrac{d\theta}{dt}\right)$
into the above equation, we get that

$$i\left(\frac{(I_3 - p_1 M_1)\cos\theta}{\lambda M\sin^2\theta} - i\frac{\cos\theta}{\sin\theta}\frac{d\theta}{dt}\right)M\sin\theta = M\cos\theta\frac{d\theta}{dt} - i\frac{d\varphi}{dt}M\sin\theta$$

and therefore,

$$\frac{d\varphi}{dt} = -\frac{(I_3 - p_1 M_1)\cos\theta}{\lambda M\sin^2\theta}\ .$$

The last expression can be written also as

(70)
$$\frac{d\varphi}{dt} = \frac{(I_3 - p_1 M_1)(p_1 - \epsilon\lambda)}{\lambda\left(M^2 - (p_1 - \epsilon\lambda)^2\right)}\ .$$

Recall now that p_1 is an even function. Therefore, the right-hand side of equation (70) is an even function and consequently it has zero residue. It follows that

$$\varphi(t) = \int_0^t \frac{(I_3 - p_1(z)M_1)p_1(z) - \epsilon\lambda)}{\lambda(M^2 - (p_1(z) - \epsilon\lambda)^2)}dz$$

is a well-defined meromorphic function. Hence,

$$w_1(t) = M\sin\theta(t)e^{\left(i\varphi(t) + t\frac{M_1}{\lambda}\right)}$$

is a meromorphic function of complex time t.

In the literature of the heavy top, the θ is known as the nutation angle while φ is known as the precession angle ([Ar1]).

The remaining variable $w_2(t)$ is then determined through the relation

$$M^2\cos^2\theta + w_1(t)w_2(t) = M^2$$

with $w_2(t) = \dfrac{M^2 \sin^2 \theta}{w_1(t)} = M \sin \theta e^{\left(i\varphi(t)+t\frac{M_1}{\lambda_1}t\right)}$ and $z_1(t)$ and $z_2(t)$ are determined through the formulas

$$z_1(t)w_2(t) = (I_3 - p_1(t)M_1) + \frac{1}{2}u(t) \quad \text{and}$$

$$z_2(t)w_1(t) = (I_3 - p_1(t)M_1) - \frac{1}{2}u(t) \ .$$

It remains to extend the integration procedure all the way down to the underlying group G. In the case that G is the semi-direct product $\mathfrak{p} \rtimes SO_3(\mathbb{C})$, the integration procedure is completely analogous to the procedure described in [**Ju2**], and the extensions to its double cover $\mathfrak{p} \rtimes SL_2(\mathbb{C})$ are straightforward. For these reasons we will omit the details of this case and concentrate on the semi simple case instead.

As we have already remarked, there are several groups G whose Lie algebra is isomorphic to $so_4(\mathbb{C})$. We shall carry out the integration on $G = SL_2(\mathbb{C}) \times SL_2(\mathbb{C})$ the double cover of $SO_4(\mathbb{C})$, which itself is a double cover of $SO_3(\mathbb{C}) \times SO_3(\mathbb{C})$.

Let us now recall Proposition 6.1 and the Lie algebra isomorphism between $so_4(\mathbb{C})$ and $sl_2(\mathbb{C}) \times sl_2(\mathbb{C})$ described by equations (48) and (49). The Hamiltonian system

$$\frac{dL}{dt}(t) = [d\mathcal{H}, L(t)], \ \frac{dg}{dt}(t) = g(t)d\mathcal{H}(t)$$

breaks up into

$$\frac{dL_1}{dt}(t) = [\Omega_1(t), L_1(t)] \text{ and } \frac{dL_2}{dt}(t) = [\Omega_2(t), L_2(t)]$$

$$\frac{dg_1}{dt}(t) = g_1(t)\Omega_1(t) \text{ and } \frac{dg_2}{dt}(t) = g_2(t)\Omega_2(t)$$

where

1. $L = \sum_1^3 M_i A_i + p_i B_i$ in $so_4(\mathbb{C})$ corresponds to L_1 and L_2 in $sl_2(\mathbb{C})$ through the formulas $L_1 = \sum_{i=1}^{i=3} u_i E_i$ and $L_2 = \sum_{i=1}^{i=3} v_i E_i$ with

$$u_1 = \frac{1}{2}(M_1 + p_1), \ u_2 = \frac{1}{2}(M_2 + p_2), \ u_3 = \frac{1}{2}(M_3 + p_3),$$

$$v_1 = \frac{1}{2}(M_1 - p_1), \ v_2 = \frac{1}{2}(M_2 - p_2), \ v_3 = \frac{1}{2}(M_3 - p_3),$$

Thus,

$$L_1 = \frac{1}{2}\begin{pmatrix} i(M_1 + p_1) & (M_2 + p_2) + i(M_3 + p_3) \\ -(M_2 + p_2) + i(M_3 + p_3) & -i(M_1 + p_1) \end{pmatrix}$$

$$L_2 = \frac{1}{2}\begin{pmatrix} i(M_1 - p_1) & (M_2 - p_2) + i(M_3 - p_3) \\ -(M_2 - p_2) + i(M_3 - p_3) & -i(M_1 - p_1) \end{pmatrix}.$$

2.

$$dH = \frac{1}{\lambda_1} M_1 A_1 + \frac{1}{\lambda} M_2 A_2 + \frac{1}{\lambda} M_3 A_3 + B_1$$

$$\Omega_1 = \frac{1}{2} \begin{pmatrix} i(\frac{1}{\lambda_1} M_1 + 1) & \frac{1}{\lambda}(M_2 + iM_3) \\ \frac{1}{\lambda}(-M_2 + iM_3) & -i(\frac{1}{\lambda_1} M_1 + 1) \end{pmatrix}$$

$$\Omega_2 = \frac{1}{2} \begin{pmatrix} i(\frac{1}{\lambda_1} M_1 - 1) & \frac{1}{\lambda}(M_2 + iM_3) \\ \frac{1}{\lambda}(-M_2 + iM_3) & -i(\frac{1}{\lambda_1} M_1 - 1) \end{pmatrix} .$$

When we recall the earlier notations

$$z_1 = M_2 + iM_3, z_2 = M_2 - iM_3, w_1 = p_2 + ip_3, w_2 = p_2 - ip_3$$

then,

$$L_1 = \frac{1}{2} \begin{pmatrix} i(M_1 + p_1) & z_1 + w_1 \\ -z_2 - w_2 & -i(M_1 + p_1) \end{pmatrix}$$

$$L_2 = \frac{1}{2} \begin{pmatrix} i(M_1 - p_1) & z_1 - w_1 \\ -z_2 + w_2 & -i(M_1 - p_1) \end{pmatrix}$$

and similarly,

$$\Omega_1 = \frac{1}{2} \begin{pmatrix} i(\frac{1}{\lambda_1} M_1 + 1) & \frac{1}{\lambda} z_1 \\ -\frac{1}{\lambda} z_2 & -i(\frac{1}{\lambda_1} M_1 + 1) \end{pmatrix}$$

$$\Omega_2 = \frac{1}{2} \begin{pmatrix} i(\frac{1}{\lambda_1} M_1 - 1) & \frac{1}{\lambda} z_1 \\ -\frac{1}{\lambda} z_2 & -i(\frac{1}{\lambda_1} M_1 - 1) \end{pmatrix} .$$

To find the solutions of the following left-invariant time varying differential systems:

$$\frac{dg_1}{dt}(t) = g_1(t)\Omega_1(t) , \quad \text{and} \quad \frac{dg_2}{dt}(t) = g_2(t)\Omega_2(t) .$$

we proceed via the coordinates on G adapted to the symmetries of the extremal equations. Recall that along each extremal curve $(g(t), L(t))$, $g^{-1}(t)L(t)g(t)$ is constant. This means that $g_1^{-1}(t)L_1(t)g_1(t) =$ constant, and that $g_2^{-1}(t)L_2(t)g_2(t) =$ constant.

Let $\text{Trace}(L_1^2) = -K_1^2$ and $\text{Trace}(L_2^2) = -K_2^2$. Both K_1 and K_2 are constant along the Hamiltonian flow.

We shall use φ_1, φ_2, φ_3 to denote the coordinates of a point g_1 in $SL_2(\mathbb{C})$ and use ψ_1, ψ_2, ψ_3 to denote the coordinates of the second point g_2 in $SL_2(\mathbb{C})$ according to the following formulas

$$g_1 = \exp\left(\phi_1 \frac{1}{2} E_1\right) \exp\left(\phi_2 \frac{1}{2} E_2\right) \exp\left(\phi_3 \frac{1}{2} E_1\right)$$
$$g_2 = \exp\left(\psi_1 \frac{1}{2} E_1\right) \exp\left(\psi_2 \frac{1}{2} E_2\right) \exp\left(\psi_3 \frac{1}{2} E_1\right)$$

Assume now that $g_1(t)$, $g_2(t)$, $L_1(t)$, $L_2(t)$ denote an extremal curve for which neither K_1 nor K_2 are equal to zero. Let Λ_1 and Λ_2 denote the elements in $sl_2(\mathbb{C})$ that satisfy

$$g_1(t) L_1(t) g_1^{-1}(t) = \Lambda_1, \quad \text{and} \quad g_2(t) L_2(t) g_2^{-1}(t) = \Lambda_2.$$

After suitable conjugations Λ_1 and Λ_2 are $\Lambda_1 = K_1 E_1$ and $\Lambda_2 = K_2 E_1$. Recall that

$$E_1 = \begin{pmatrix} i & 0 \\ 0 & -i \end{pmatrix}, E_2 = \begin{pmatrix} 0 & 1 \\ -1 & 0 \end{pmatrix}, E_3 = \begin{pmatrix} 0 & i \\ i & 0 \end{pmatrix}.$$

Then,

$$L_1(t) = K_1 g_1^{-1}(t) E_1 g_1(t), \quad \text{and} \quad L_2(t) = K_2 g_2^{-1}(t) E_1 g_2(t)$$

and therefore,

$$L_1(t) = K_1 e^{-\frac{1}{2} E_1 \varphi_3} e^{-\frac{1}{2} E_2 \varphi_2} E_1 e^{\frac{1}{2} E_2 \varphi_2} e^{\frac{1}{2} E_1 \varphi_3},$$
$$L_2(t) = K_2 e^{-\frac{1}{2} E_1 \psi_3} e^{-\frac{1}{2} E_2 \psi_2} E_1 e^{\frac{1}{2} E_2 \psi_2} e^{\frac{1}{2} E_1 \psi_3}.$$

It follows that

$$L_1(t) = i K_1 \begin{pmatrix} \cos\varphi_2 & e^{-i\varphi_3}\sin\varphi_2 \\ e^{i\varphi_3}\sin\varphi_2 & -\cos\varphi_2 \end{pmatrix}, \quad \text{and}$$
$$L_2(t) = i K_2 \begin{pmatrix} \cos\psi_3 & e^{-i\psi_3}\sin\psi_2 \\ e^{i\psi_3}\sin\psi_2 & -\cos\psi_2 \end{pmatrix}$$

and therefore, $(M_1 + p_1) = K_1 \cos\varphi_2$, $(M_1 - p_1) = K_2 \cos\psi_2$.

Furthermore,

(70) $\qquad z_1 + w_1 = i K_1 e^{-i\varphi_3}\sin\varphi_2, \quad z_2 + w_2 = -i K_1 e^{i\varphi_3}\sin\varphi_2,$

and

$$z_1 - w_1 = i K_2 e^{-i\psi_3}\sin\varphi_2, z_2 - w_2 = -i K_2 e^{i\psi_3}\sin\varphi_3.$$

The remaining variables φ_1 and ψ_1 are obtained as follows: Since $g_1(t) = e^{\frac{1}{2} E_1 \varphi_1} e^{\frac{1}{2} E_2 \varphi_2} e^{\frac{1}{2} E_1 \varphi_3}$,

$$g_1^{-1}\frac{dg_1}{dt} = \frac{1}{2}\left(\dot{\varphi_1} g_1^{-1} E_1 g_1 + \dot{\varphi_2}\left(e^{-\frac{1}{2} E_1 \varphi_3} E_2 e^{\frac{1}{2} E_1 \varphi_3}\right) + \dot{\varphi_3} E_1\right)$$

But $\frac{1}{2}g^{-1}(t)E_1 g_1(t) = \frac{1}{K_1}L_1(t)$, and

$$e^{-\frac{1}{2}E_1\varphi_3}\frac{1}{2}E_2 e^{\frac{1}{2}E_1\varphi_3} = \frac{1}{2}\begin{pmatrix} 0 & e^{i\varphi_3} \\ -e^{-i\varphi_3} & 0 \end{pmatrix}.$$

Hence,

$$g_1^{-1}\frac{dg_1}{dt} = i\frac{\dot\varphi_1(t)}{K_1}L_1(t) + \frac{1}{2}(\dot\varphi_2\begin{pmatrix} 0 & e^{-i\varphi_3} \\ -e^{i\varphi_3} & 0 \end{pmatrix} + \frac{\dot\varphi_3}{2}\begin{pmatrix} i & 0 \\ 0 & -i \end{pmatrix})) = d\mathcal{H}_1 .$$

where $d\mathcal{H}_1 = \dfrac{1}{2}\begin{pmatrix} i\left(\dfrac{M_1}{\lambda_1}+1\right) & \dfrac{z_1}{\lambda} \\ -\dfrac{z_2}{\lambda} & -i\left(\dfrac{M_1}{\lambda_1}+1\right) \end{pmatrix}.$

The equation

$$\frac{\dot\varphi_1}{2}\begin{pmatrix} \cos\varphi_2 & e^{-i\varphi_3}\sin\varphi_2 \\ e^{i\varphi_3}\sin\varphi_2 & -\cos\varphi_2 \end{pmatrix} + \frac{1}{2}\dot\varphi_2\begin{pmatrix} 0 & e^{-i\varphi_3} \\ -e^{i\varphi_3} & 0 \end{pmatrix} + \frac{\dot\varphi_3}{2}\begin{pmatrix} 1 & 0 \\ 0 & -1 \end{pmatrix} = d\mathcal{H}_1$$

yields

$$\dot\varphi_1 e^{-i\varphi_3}\sin\varphi_2 + \dot\varphi_2 e^{-i\varphi_3} = \frac{1}{\lambda}z_1$$

$$\dot\varphi_1 e^{i\varphi_3}\sin\varphi_2 - \dot\varphi_2 e^{i\varphi_3} = -\frac{1}{\lambda}z_2 .$$

Therefore,

$$\dot\varphi_1 \sin\varphi_2 = \frac{1}{\lambda}(e^{i\varphi_3}z_1 - e^{-i\varphi_3}z_2),$$

or

$$\dot\varphi_1 = \frac{1}{\lambda\sin\varphi_2}\left(\frac{z_1}{e^{-i\varphi_3}} - \frac{z_2}{e^{i\varphi_3}}\right).$$

It follows from (70) that $e^{-i\varphi_3}\sin\varphi_2 = \dfrac{z_1 + w_1}{iK_1}$ and $-e^{i\varphi_3}\sin\varphi_2 = \dfrac{z_2 + w_2}{iK_1}$. Then,

$$\dot\varphi_1 = i\frac{K_1}{2\lambda}\left(\frac{z_1}{z_1 + w_1} + \frac{z_2}{z_2 + w_2}\right) = i\frac{K_1}{2\lambda}\left(\frac{2z_1 z_2 + z_1 w_2 - z_2 w_1}{(z_1 + w_1)(z_2 + w_2)}\right).$$

The constants of motion $I_1 = \dfrac{z_1 z_2}{2\lambda} + p_1$ $I_2 = p_1 M_1 + \dfrac{1}{2}(z_1 w_1 + z_1 w_2)$ and $I_3 = p_1^2 + z_1 z_2 + z_1 w_2 + z_2 w_1$ imply that

$$\dot\varphi_1 = i\frac{K_1}{2\lambda}\left(\frac{4\lambda(I_1 - p_1) + 2(I_2 - p_1 M_1)}{2\lambda(I_2 - p_1 M_1) + I_3 - p_1^2}\right).$$

The right-hand side of the above equation is an even function (since p_1 is even) and hence, is residue free. Therefore, $\varphi_1(t)$ is single valued. An analogous argument applied to $g_2^{-1}(t)\dfrac{dg_2}{dt} = d\mathcal{H}_2$ yields that $\psi_1(t)$ is also single valued. This completes our integration procedure.

Apart from providing explicit solutions, this procedure shows that the solutions are meromorphic functions of complex time on the entire cotangent bundle of G.

X. Elastic Curves for the case of Kowalewski

We now concentrate on the case $\lambda = \lambda_1 = \lambda_2 = 2\lambda_3$ and $b_3 = 0$. As shown earlier (equation (65))

$$I_4 = \left(z_1^2 - \lambda b(w_1 - \epsilon \lambda b)\right)\left(z_2^2 - \lambda \bar{b}(w_2 - \epsilon \lambda \bar{b})\right)$$

is an integral of motion for the Hamiltonian

$$\mathcal{H} = \frac{1}{2\lambda}(M_1^2 + M_2^2) + \frac{1}{2\lambda_3}M_3^2 + b_1 p_1 + b_2 p_2$$

where $z_1 = \frac{1}{2}(M_1 + iM_2)$, $z_2 = \frac{1}{2}(M_1 - iM_2)$, $w_1 = p_1 + ip_2$, $w_2 = p_1 - ip_2$, $b = b_1 + ib_2$, and where $\bar{b} = b_1 - ib_2$.

In contrast to the case of Lagrange, where the extra integral of motion is due to the symmetry of the problem and occurs as a simple consequence of Noether's theorem, the integral of motion discovered by Kowalewski has always been shrouded in mathematical mystery. Our first aim is to provide geometric explanations for the mysterious relations $\lambda_1 = \lambda_2 = 2\lambda_3$, and $\alpha_3 = 0$. In doing so we will also provide simple explanations for the existence of extra integrals of motion discovered by A.I. Bobenko, A.G. Reyman and M.A. Semenov-Tian-Shansky in a Hamiltonian system called the Kowalewski gyrostat in two constant fields ([**Bb**]).

1. Kowalewski gyrostat in two constant fields.

The Hamiltonian that corresponds to the Kowalewski gyrostat in two constant fields contains an extra parameter γ and is of the form

$$\mathcal{H} = \frac{1}{2}(M_1^2 + M_2^2 + 2M_3^2 + 2M_3\gamma) - g_1 - h_2 \ .$$

The authors of ([**Bb**]) obtain additional integrals of motion for the above Hamiltonian through a Lax-pair representation on $sp_4(\mathbb{R})$ in which the integral of motion discovered by S. Kowalewski appears as a particular case when $\gamma = h_2 = 0$. We reproduce the results of ([**Bb**]) by showing that the above integrals of motion are a consequence of certain symmetries inherited from $so_5(\mathbb{C})$. The exposition is based on a brilliant observation of A. Savu (personally communicated, and subsequently reported in [**IvS**]) that the parameter γ itself is an integral of motion in $sp_4(\mathbb{R})$.

To explain these symmetries in detail, we shall consider $SO_5(\mathbb{C})$ together with an involutive automorphism σ on $SO_5(\mathbb{C})$ given by $\sigma(g) = JgJ^{-1}$ where J is a diagonal matrix with the diagonal entries -1, -1, 1, 1, 1.

Then the tangent map σ_* at the identity induces a splitting of $so_5(\mathbb{C})$ into the direct sum $\mathfrak{k} \oplus \mathfrak{p}$ where $\mathfrak{k}$ denotes the Lie subalgebra of fixed points of σ_*. Matrices M in $so_5(\mathbb{C})$ will be written in block form $M = \begin{pmatrix} A & -B^T \\ B & C \end{pmatrix}$ where A is a 2×2 block, B a 3×2 block, C a 3×3 block, and where B^T the denotes the transpose of B. We shall write $M_{\mathfrak{k}}$ for the projection of M on $\mathfrak{k}$, and $M_{\mathfrak{p}}$ for the projection of M on $\mathfrak{p}$. It is easily verified that matrices $M_{\mathfrak{k}}$ are of the form $\begin{pmatrix} A & 0 \\ 0 & C \end{pmatrix}$, while matrices $M_{\mathfrak{p}}$ are of the form $= \begin{pmatrix} 0 & -B^T \\ B & 0 \end{pmatrix}$.

It follows that $\mathfrak{k}$ is a 4-dimensional sub algebra of $so_5(\mathbb{C})$, while $\mathfrak{p}$ is a 6-dimensional vector subspace of $so_5(\mathbb{C})$. Furthermore, $\mathfrak{k} = \mathfrak{k}_1 \oplus \mathfrak{k}_2$ with $\mathfrak{k}_1$ isomorphic to $so_3(\mathbb{C})$ and $\mathfrak{k}_1$ isomorphic to $so_2(\mathbb{C})$ i.e., $\begin{pmatrix} A & 0 \\ 0 & 0 \end{pmatrix} \in \mathfrak{k}_2$ while $\begin{pmatrix} 0 & 0 \\ 0 & C \end{pmatrix} \in \mathfrak{k}_1$.

The vector space $\mathfrak{p}$ is equal to $V_1 \oplus V_2$ where V_1 and V_2 are the vector spaces of matrices M_p having the second column (respectively the first column) of B equal to zero. Evidently, V_1 and V_2 are three dimensional complex vector spaces. In addition to the usual Cartan decomposition

$$[\mathfrak{p}, \mathfrak{p}] \subseteq \mathfrak{k}, [\mathfrak{p}, \mathfrak{k}] \subseteq \mathfrak{p}, [\mathfrak{k}, \mathfrak{k}] = \mathfrak{k}$$

the following Lie algebraic conditions also hold:

(71)
$$[\mathfrak{k}_1, \mathfrak{k}_2] = 0 \, , \; [\mathfrak{k}_1, V_1] \subseteq V_2 \, , \; [\mathfrak{k}_2, V_2] \subseteq V_1 \, , \; [V_1, V_2] \subseteq \mathfrak{k}_2 \, , \; [V_1, V_1] \subseteq \mathfrak{k}_1 \, , \; [V_2, V_2] \subseteq \mathfrak{k}_1 \, .$$

We shall use A_1, A_2, A_3, A_4 to denote a basis in $\mathfrak{k}$ where A_1, A_2, A_3 denote the standard basis in $so_3(\mathbb{C})$ embedded in $\mathfrak{k}_1$, while $A_4 = \begin{pmatrix} A & 0 \\ 0 & 0 \end{pmatrix}$, with $A = \begin{pmatrix} 0 & -1 \\ 1 & 0 \end{pmatrix}$ is in $\mathfrak{k}_2$.

Let B_1, B_2, B_3 denote the standard basis for V_1, and let C_1, C_2, C_3 denote the standard basis for V_2, i.e., the bases which coincide with the standard basis e_1, e_2, e_3 in $\mathbb{C}^3$ under the usual identification of each of V_1 and V_2 with $\mathbb{C}^3$.

We regard $so_5^*(\mathbb{C})$ as the direct sum $\mathfrak{k}_1^* \oplus \mathfrak{k}_2^* \oplus V_1^* + V_2^*$, in which case M_1, M_2, M_3, M_4, g_1, g_2, g_3, h_1, h_2, h_3 will denote the coordinates of a point ℓ in $so_5^*(\mathbb{C})$ relative to the dual basis A_1^*, A_2^*, A_3^*, A_4^*, B_1^*, B_2^*, B_3^*, C_1^*, C_2^*, C_3^*.

We shall now consider the Hamiltonian equations in $T^*SO_5(\mathbb{C})$ induced by

$$\mathcal{H} = \frac{1}{2\lambda}(M_1^2 + M_2^2) + \frac{1}{2\lambda_3}M_3^2 + \alpha M_3 + b_1 g_1 + b_2 g_2 + b_3 g_3 + c_1 h_1 + c_2 h_2 + c_3 h_3 \; ,$$

which we will simply write as

$$\mathcal{H} = \frac{1}{2\lambda}(M_1^2 + M_2^2) + \frac{1}{2\lambda_3}M_3^2 + \alpha M_3 + b \cdot h + c \cdot g \; .$$

Since $SO_5(\mathbb{C})$ is semi-simple, the Hamiltonian equations can be written as

(72) $$\frac{dg(t)}{dt} = g(t)(d\mathcal{H}) \; , \quad \frac{dL}{dt} = [d\mathcal{H}, L]$$

where

1. $d\mathcal{H} = \Omega + \alpha A_3 + B + C$ with $B = \sum_{i=1}^{3} b_i B_i$, $C = \sum_{i=1}^{3} c_i C_i$, and

$$\Omega = \frac{1}{\lambda}M_1 A_1 + \frac{1}{\lambda}M_2 A_2 + \frac{1}{\lambda_3}M_3 A_3,$$

and

2. $L = M + L_4 + g + h$ with $M = \sum_{i=1}^{3} M_i A_i, L_4 = M_4 A_4, g = \sum_{i=1}^{3} g_i B_i$ and $h = \sum_{i=1}^{3} h_i C_i.$

Then, equation $\dfrac{dL}{dt} = [d\mathcal{H}, L]$ becomes,

$$\frac{dM}{dt} = [\Omega, M] + \alpha[A_3, M] + [B, g] + [C, h]$$
$$\frac{dL_4}{dt} = [B, h] + [C, g]$$
$$\frac{dg}{dt} = [\Omega, g] + [B, M] + \alpha[A_3, g] + [C, L_4]$$
$$\frac{dh}{dt} = [\Omega, h] + \alpha[A_3, h] + [C, M] + [B, L_4],$$

as a consequence of the Lie bracket structure described by (71).

The preceding equations may be also written as

(73)
$$\frac{d\hat{M}}{dt} = \hat{M} \times \hat{\Omega} + \alpha\hat{M} \times \hat{A}_3 + \hat{g} \times \hat{B} + \hat{h} \times \hat{C}$$

$$\frac{dM_4}{dt} = -\hat{h} \cdot \hat{B} + \hat{g} \cdot \hat{C}$$

$$\frac{d\hat{g}}{dt} = \hat{g} \times \hat{\Omega} + \hat{M} \times \hat{B} + \alpha(g \times \hat{A}_3) - M_4\hat{C}$$

$$\frac{d\hat{h}}{dt} = \hat{h} \times \hat{\Omega} + \hat{M} \times \hat{C} + \alpha(\hat{h} \times \hat{A}_3) + M_4\hat{B}$$

where $\hat{M} = \begin{pmatrix} M_1 \\ M_2 \\ M_3 \end{pmatrix}$, $\hat{g} = \begin{pmatrix} g_1 \\ g_2 \\ g_3 \end{pmatrix}$, $\hat{h} = \begin{pmatrix} h_1 \\ h_2 \\ h_3 \end{pmatrix}$, $\hat{B} = \begin{pmatrix} b_1 \\ b_2 \\ b_3 \end{pmatrix}$, $\hat{C} = \begin{pmatrix} c_1 \\ c_2 \\ c_3 \end{pmatrix}$.

The products are the usual inner products on $\mathbb{C}^3$, i.e., $\hat{h} \cdot \hat{B} = \sum_{i=1}^{3} h_i b_i$ and $\hat{g} \cdot \hat{C} = \sum_{i=1}^{3} c_i g_i$.

It follows from (73) that

$$\frac{dM_3}{dt} = g_1 b_2 - g_2 b_1 + h_1 c_2 - h_2 c_1$$

and therefore,

$$\frac{d}{dt}(M_3 + M_4) = g_1 b_2 - g_2 b_1 + h_1 c_2 - h_2 c_1 - h_1 b_1 - h_2 b_2 - h_3 b_3 + g_1 c_1 + g_2 c_2 + g_3 c_3$$

$$= g_1(b_2 + c_1) + g_2(c_2 - b_1) - h_1(b_1 - c_2) - h_2(b_2 + c_1) + g_3 c_3 - h_3 b_3 .$$

Hence, $\frac{d}{dt}\big(M_3(t) + M_4(t)\big) = 0$ whenever $c_1 + b_2 = 0$, $c_2 - b_1 = 0$, and $c_3 = b_3 = 0$. That is, $M_3(t) + M_4(t)$ is an integral of motion for system (73) whenever $\hat{B} = \begin{pmatrix} b_1 \\ b_2 \\ 0 \end{pmatrix}$, and $\hat{C} = \hat{B}^\perp = \begin{pmatrix} -b_2 \\ b_1 \\ 0 \end{pmatrix}$.

Let $M_3(t) + M_4 = \gamma$ and let $\hat{B} = \begin{pmatrix} b_1 \\ b_2 \\ 0 \end{pmatrix}$, and $\hat{C} = \hat{B}^\perp = \begin{pmatrix} -b_2 \\ b_1 \\ 0 \end{pmatrix}$.

Then equations (73) can be written as:

(74)
$$\frac{d\hat{g}}{dt} = \hat{g} \times \hat{\Omega} + \hat{M} \times \hat{B} + \alpha(\hat{g} \times A_3) - (\gamma - M_3)\hat{B}^\perp ,$$

$$\frac{d\hat{h}}{dt} = \hat{h} \times \hat{\Omega} + \hat{M} \times \hat{B}^\perp + \alpha(\hat{h} \times A_0) - (\gamma - M_3)\hat{B} .$$

Now normalize λ to $\lambda = 1$ and let $\lambda_3 = \dfrac{1}{2}$, i.e., assume that the Kowalewski conditions $\lambda_1 = \lambda_2 = 2\lambda_3$ hold. Then,

$$\hat{B} \times \hat{\Omega} + \hat{M} \times \hat{B} = \hat{B} \times (\hat{\Omega} - \hat{B}) = \hat{B} \times (M_3 e_3) = -M_3 \hat{B}^\perp \, , \quad \text{which implies that}$$

$$\hat{B} \times \hat{\Omega} + \hat{M} \times \hat{B} + M_3 \hat{B}^\perp = 0 \, .$$

Similarly,

$$\hat{B}^\perp \times \hat{\Omega} + \hat{M} \times \hat{B} - M_3 \hat{B} = 0 \, .$$

Consider now the transformation

$$g = \bar{g} + B \quad \text{and} \quad h = \bar{h} + B^\perp \, .$$

Then,

$$\frac{d\bar{g}}{dt} = \frac{dg}{dt} = [\Omega, \bar{g} + B] + [B, M] + \alpha[A_3, \bar{g} + B] + M_3 B^\perp - r B^\perp$$
$$= [\Omega, \bar{g}] + \alpha[A_3, \bar{g}] - \gamma B^\perp + \alpha[A_3, B]$$

because $[\Omega, B] + [B, M] + M_3 B^\perp = 0$.

Similarly,

$$\frac{d\bar{h}}{dt} = [\Omega, \bar{h} + B^\perp] + [B^\perp, M] + \alpha[A_3, \bar{h} + B^\perp] - M_3 B + \gamma B$$
$$= [\Omega, \bar{h}] + \alpha[A_3, \bar{h}] + \alpha[A_3, B^\perp] + \gamma B \, .$$

Since $\alpha[A_3, B^\perp] + \gamma B = (\alpha + r)B$, and $\alpha[A_3, B] - r B^\perp = -(\alpha + \gamma)B^\perp$ the preceding equations reduce to

$$(75) \qquad \frac{d\bar{g}}{dt} = [\Omega, \bar{g}] - \gamma[A_3, \bar{g}] \, , \quad \text{and} \quad \frac{d\bar{g}}{dt} = [\Omega, \bar{h}] - \gamma[A_3, \bar{h}]$$

whenever $\alpha = -\gamma$. Together with

$$(76) \qquad \frac{dM}{dt} = [\Omega, M] + [B, \bar{g}] + [B^\perp, \bar{h}] + \gamma[A_3, M]$$

these equations coincide with the Hamiltonian equations on the semi-direct product $(\mathbb{C}^3 \oplus \mathbb{C}_3^3) \rtimes SO_3(\mathbb{C})$ corresponding to the Hamiltonian

$$\mathcal{H} = \frac{1}{2}(M_1^2 + M_2^2 + 2M_3^2) + b_1 h_1 + b_2 h_2 - b_2 g_1 + b_1 g_2 - \gamma M_3 \, .$$

This Hamiltonian function considered as a function on the dual of $(\mathbb{R}^3 + \mathbb{R}^3) \rtimes SO_3(\mathbb{R})$ is called the Kowalewski gyrostat in two constant fields by the authors in ([**Bb**]).

Now turn to the integrals of motion and to their Lax-pair representation. The Hamiltonian equations in $so_5(\mathbb{C})$

$$\frac{dL}{dt} = [d\mathcal{H}, L]$$

admit two integrals of motion given by

$$\langle L, L\rangle = -\frac{1}{2}\text{Trace}(L^2)\ , \quad \text{and } \langle L^2, L^2\rangle = -\frac{1}{2}\text{Trace}(L^4)$$

as we have demonstrated before. It is transparent that equations (75) yield three additional integrals of motion:

$$I_2 = \bar{\hat{h}} \cdot \bar{\hat{h}} = \bar{h}_1^2 + \bar{h}_2^2 + \bar{h}_3^2$$

$$I_3 = \bar{\hat{g}} \cdot \bar{\hat{g}} = \bar{g}_1^2 + \bar{g}_2^2 + \bar{g}_3^3$$

$$I_4 = \hat{g} \cdot \hat{h} = \hat{g}_1\hat{h}_1 + \hat{g}_2\hat{h}_2 + \hat{g}_3\hat{h}_3$$

Furthermore,

$$\langle L, L\rangle = M_1^2 + M_2^2 + M_3^2 + M_4^2 + g_1^2 + g_2^2 + g_3^2 + h_1^2 + h_2^2 + h_3^2\ ,$$

which along the integral curves of $\frac{dL}{dt} = [d\mathcal{H}, L]$ reduces to

$$\langle L(t), L(t)\rangle = M_1^2 + M_2^2 + M_3^2 + (\gamma - M_3)^2 + 2\bar{\hat{h}} \cdot \hat{B} + 2\hat{\bar{h}} \cdot \hat{B}^\perp$$
$$+ \bar{\hat{h}} \cdot \bar{\hat{h}} + \bar{\hat{g}} \cdot \bar{\hat{g}} + \hat{B} \cdot \hat{B} + \hat{B}^\perp \cdot \hat{B}^\perp$$
$$= 2\mathcal{H} + \bar{\hat{h}} \cdot \bar{\hat{h}} + \bar{\hat{g}} \cdot \bar{\hat{g}} + \hat{B} \cdot \hat{B} + \hat{B}^\perp \cdot \hat{B}^\perp\ .$$

It follows that Hamiltonian $\mathcal{H} = I_1$, the integrals of motion I_2, I_3 and the Casimir function $\langle L, L\rangle$ are functionally dependent.

We demonstrate that $\langle L^2, L^2\rangle$ produces two extra integrals of motion of which one will yield the integral of Kowalewski as reported in ([**Bb**]). Their existence ir rooted in an additional symmetry already used in the classification of the meromorphic cases.

The equations

(75)
$$\frac{dM}{dt} = [\Omega, M] + [B, \bar{g}] + [B^\perp, \bar{h}] - \gamma[A_3, M]$$
$$\frac{d\bar{g}}{dt} = [\Omega, \bar{g}] - \gamma[A_3, \bar{g}]\ , \quad \frac{d\bar{h}}{dt} = [\Omega, \bar{h}] - \gamma[A_3, \bar{h}]$$

are invariant under the dilations $\bar{h} \rightarrow \frac{1}{\lambda}\bar{h}$ $\bar{g} \rightarrow \frac{1}{\lambda}\bar{g}$, $B \rightarrow \lambda B$ and $B^\perp \rightarrow \lambda B^\perp$.

Hence, the traces of

$$L(\lambda) = M + L_4 + g + h = M + (\gamma - M_3)A_4 + \frac{1}{\lambda}\bar{g} + \lambda B + \frac{1}{\lambda}\bar{h} + \lambda B^\perp$$

must be the integrals of motion for the above Hamiltonian system for each λ.

Let us now investigate the integrals of motion associated with $I = \frac{1}{2}\big(\langle L(\lambda)^2, L(\lambda)^2\rangle - \langle L(\lambda, L(\lambda))\rangle\big).$

An easy calculation yields that

$$I(\lambda) = \big(h(\lambda) \cdot h(\lambda)\big)\big(g(\lambda) \cdot g(\lambda)\big) - \big(g(\lambda) \cdot h(\lambda)\big)^2 + \big(\hat{M} \cdot h(\lambda)\big)^2 + \big(\hat{M} \cdot g(\lambda)\big)^2$$
$$- 2(\gamma - M_3)h(\lambda) \cdot Mg(\lambda) + (\gamma - M_3)^2 \hat{M} \cdot \hat{M}$$

with $h(\lambda) = \dfrac{1}{\lambda}\bar{h} + \lambda B$, and $g(\lambda) = \dfrac{1}{\lambda}\bar{g} + \lambda B^{\perp}$.

Hence,

(76)
$$I(\gamma, \lambda) = \lambda^4 I_5 + \lambda^2 I_6 + I_7 + \frac{1}{\lambda^2}I_8 + \frac{1}{\lambda 4}I_9$$

and each of I_5, I_6, I_7, I_8, I_9 is an integral of motion for $\mathcal{H}$. It is easy to see that $I_5 = 0$ because $\hat{B} \cdot \hat{B}^{\perp} = 0$. It is also easy to verify that

$$I_9 = (\bar{h} \cdot \bar{h})(\bar{g} \cdot \bar{g}) - (\bar{h} \cdot \bar{g})^2 = I_2 I_3 - I_4^2 \ .$$

Hence, I_9 is not a new integral of motion. The same applies to I_6 for the following reasons:

$$I_6 = (\bar{h} \cdot \hat{B}^{\perp})\hat{B} \cdot \hat{B} + (\bar{g} \cdot \hat{B})\hat{B}^{\perp} \cdot \hat{B}^{\perp} + (\hat{M} \cdot \hat{B})^2 + (\hat{M} \cdot \hat{B}^{\perp})^2 - 2(\gamma_3 - M_3)(\hat{B}^{\perp} \cdot M\hat{B}) \ .$$

A simple calculation shows that $\hat{B} \cdot M\hat{B} = M_3 \hat{B} \cdot \hat{B}$ and that $(\hat{M} \cdot \hat{B})^2 - (\hat{M} \cdot \hat{B}^{\perp})^2 = (M_1^2 + M_2^2)\hat{B} \cdot \hat{B}$. Hence,

$$I_6 = (\hat{B} \cdot \hat{B})(\bar{h} \cdot \hat{B}^{\perp} + \bar{g} \cdot \hat{B} + M_1^2 + M_2^2 + 2M_3^2 - 2\gamma_3 M_3) = 2(\hat{B} \cdot \hat{B})\mathcal{H} \ .$$

We show that the remaining quantities I_7 and I_8 produce new integrals of motion, which in the particular cases $b_1 = -1$, $b_2 = 0$ agree with the ones in [**Bb**].

It follows that

$$I_8 = 2(\bar{g} \cdot \bar{g})(\bar{h} \cdot \hat{B}^{\perp}) + 2(\bar{h} \cdot \bar{h})(\bar{g} \cdot B) - 2(\bar{g} \cdot \bar{h})(\bar{g} \cdot \hat{B}^{\perp} + \bar{h} \cdot \hat{B})$$
$$+ (\hat{M} \cdot \bar{h})^2 + (\hat{M} \cdot \bar{h})^2 - 2(\gamma - M_3)\bar{h} \cdot M\bar{h} \ ,$$

and this integral of motion is the same as the one reported in [**Bb**] (their parameter γ is the negative of the one used in this paper, and their choice of $\hat{B}$ is $b_1 = -1$ $b_2 = 0$).

It remains to calculate I_7. It follows that

$$I_7 = (\bar{h} \cdot \bar{h})(\hat{B} \cdot \hat{B}) + (\bar{g} \cdot \bar{g})(\hat{B}^{\perp} \cdot \hat{B}^{\perp}) + 4(\bar{h} \cdot \hat{B}^{\perp})(\bar{g} \cdot \hat{B}^{\perp}) - (\bar{h} \cdot \hat{B}^{\perp} + g \cdot \hat{B})^2$$
$$+ 2(\hat{M} \cdot \bar{h})(\hat{M} \cdot \hat{B}^{\perp}) + 2(\hat{M} \cdot \bar{g})(\hat{M} \cdot \hat{B}) - 2(\gamma - M_3)(\bar{h} \cdot M\hat{B} + \hat{B}^{\perp} \cdot M\bar{g})$$
$$+ (\gamma - M_3)^2(M_1^2 + M_2^2 + M_3^2) \ .$$

Since $I_2 = \bar{h} \cdot \bar{h}$ and $I_3 = \bar{g} \cdot \bar{g}$ the reduced expression $I_7 - (\bar{h} \cdot \bar{h})(\hat{B} \cdot \hat{B}) - (\bar{g} \cdot \bar{g})(\hat{B}^\perp \cdot \hat{B}^\perp)$ is also an integral of motion. To further simplify, it will be necessary to make several side calculations. To begin with

$$\bar{h} \cdot M\hat{B} + \hat{B}^\perp \cdot Mg = \bar{h} \cdot M\hat{B} - g \cdot M\hat{B}^\perp = M_3(\bar{g} \cdot \hat{B} + \bar{h} \cdot \hat{B})$$
$$+ h_3(M_1 b_2 - M_2 b_1) - g_3(M_2 b_2 + M_1 b_1) \,,$$

and

$$(\gamma - M_3)^2 M_3^2 - 2(\gamma - M_3)M_3(\bar{g} \cdot \hat{B} + \bar{h} \cdot \hat{B})$$
$$= (\bar{g} \cdot \hat{B} + \bar{h} \cdot \hat{B}^\perp - (\gamma - M_3)M_3)^2 - (\bar{g} \cdot \hat{B} + \bar{h} \cdot \hat{B}^\perp)^2$$
$$= \left(\mathcal{H} - \frac{1}{2}(M_1^2 + M_2^2)\right)^2 - (\bar{g} \cdot \hat{B} + \bar{h} \cdot \hat{B}^\perp)^2 \,.$$

Then,

$$(\gamma - M_3)^2(M_1^2 + M_2^2 + M_3^2) - 2(\gamma - M_3)(\bar{h} \cdot M\hat{B} + \hat{B}^\perp \cdot Mg)$$
$$= (\gamma - M_3)^2(M_1^2 + M_2^2) + (\gamma - M_3)^2 M_3^2 - 2(\gamma - M_3)M_3(\bar{h} \cdot \hat{B}^\perp + \bar{g} \cdot \hat{B})$$
$$- 2(\gamma - M_3)(h_3(M_1 b_2 - M_2 b_1) - g_3(M_2 b_2 + M_1 b_1)$$
$$= (\gamma - M_3)^2(M_1^2 + M_2^2) + \left(\mathcal{H} - \frac{1}{2}(M_1^2 + M_2^2)\right)^2 - (\bar{g} \cdot \hat{B} + \bar{h} \cdot \hat{B}^\perp)^2$$
$$- 2(\gamma - M_3)\big(h_3(M_1 b_2 - M_2 b_1) - g_3(M_2 b_2 + M_1 b_1)\big) \,.$$

Since

$$-\gamma M_3 + M_3^2 = \mathcal{H} - (\bar{h} \cdot \hat{B}^\perp + \bar{g} \cdot \hat{B}) - \frac{1}{2}(M_1^2 + M_2^2)$$

the preceeding expression can be written as

$$(\gamma^2 - \gamma M_3)(M_1^2 + M_2^2) + \left(\mathcal{H} - (\bar{g} \cdot \hat{B} + \bar{h} \cdot \hat{B}^\perp) - \frac{1}{2}(M_1^2 + M_2^2)\right)(M_1^2 + M_2^2)$$
$$+ \left(\mathcal{H} - \frac{1}{2}(M_1^2 + M_2^2)\right)^2 - (g \cdot \hat{B} + \bar{h} \cdot \hat{B}^\perp)^2$$
$$- 2(\gamma - M_3)\big(h_3(M_1 b_2 - M_2 b_1) - g_3(M_2 b_2 + M_1 b_1)\big)$$
$$= \gamma(\gamma - M_3)(M_1^2 + M_2^2) + \mathcal{H}^2 - \frac{1}{4}(M_1^2 + M_2^2) - (M_1^2 + M_2^2)(g \cdot \hat{B} + \bar{h} \cdot \hat{B}^\perp)$$
$$- (\bar{g} \cdot \hat{B} + \bar{h} \cdot \hat{B}^\perp)^2 \,.$$

Let

$$J = I_7 - (\bar{h} \cdot \bar{h})(\hat{B} \cdot \hat{B}) - (\bar{g} \cdot \bar{g})(\hat{B}^\perp \cdot \hat{B}^\perp) - \mathcal{H}^2.$$

It then follows from the preceeding calculations that

$$J = 4(\bar{h} \cdot \hat{B}^\perp)(\bar{g} \cdot \hat{B}) - (\bar{h} \cdot \hat{B} + \bar{g} \cdot \hat{B}^\perp)^2 + 2(M_1 \bar{h}_1 + M_2 \bar{h}_2)(\hat{M} \cdot \hat{B}^\perp)$$
$$+ 2(M_1 \bar{g}_1 + M_2 \bar{g}_2)(\hat{M} \cdot \hat{B}) - \frac{1}{4}(M_1^2 + M_2^2)^2 - (M_1^2 + M_2^2)(\bar{g}\hat{B} + \bar{h}\hat{B}^\perp)$$
$$- (\bar{g} \cdot \hat{B} + \bar{h} \cdot \hat{B}^\perp)^2 + \gamma(\gamma - M_3)(M_1^2 + M_2^2)$$
$$- 2\gamma\big(h_3(M_1 b_1 - M_2 b_1) - g_3(M_2 b_2 + M_1 b_1)\big) \ .$$

Let J_0 denote the part of J which is independent of γ. It follows that

$$J = J_0 + \gamma\big((\gamma + M_3)(M_1^2 + M_2^2) - 2\bar{h}_3(M_1 b_2 - M_2 b_1) + 2\bar{g}_3(M_2 b_2 + M_1 b_1)\big) \ .$$

It remains to relate J_0 to the integral of motion obtained by S. Kowalewski. To begin with $4(\bar{h} \cdot \hat{B}^\perp)(\bar{g} \cdot \hat{B}) - (\bar{g} \cdot \hat{B}^\perp + \bar{h} \cdot \hat{B}) - (\bar{g} \cdot \hat{B} + h \cdot \hat{B}^\perp)^2 = -(\hat{B} \cdot \hat{B})(\bar{h}_1 - \bar{g}_2)^2 + (\bar{h}_2 - \bar{g}_1)^2)$, and hence,

$$J_0 = -\frac{1}{4}(M_1^2 + M_2^2)^2 - (M_1^2 + M_2^2)(\bar{g}\hat{B} + \bar{h} \cdot \hat{B}^\perp)$$
$$+ 2(M_1 \bar{h}_1 + M_2 \bar{h}_2)(\hat{M} \cdot \hat{B}^\perp) + 2(M_1 \bar{g}_1 + M_2 \bar{g}_2)(\hat{M} \cdot \hat{B})$$
$$+ (\hat{B} \cdot \hat{B})\big((\bar{h}_1 + \bar{g}_2)^2 + (\bar{h}_2 - \bar{g}_1)^2\big) \ .$$

Reintroduce the notations used earlier: $z_1 = \frac{1}{2}(M_1 + iM_2)$, $z_2 = \frac{1}{2}(M_1 - iM_2)$, $w_1 = (\bar{h}_1 + \bar{g}_2) + i(\bar{h}_2 - \bar{g}_1)$, $w_2 = (\bar{h}_1 + \bar{g}_2) - i(\bar{h}_2 - \bar{g}_1)$, $c_1 = i(b_1 + ib_2)$ and $c_2 = -i(b_1 - ib_2)$.

An easy calculation, which we leave to the reader, shows that

$$-(M_1^2 + M_2^2)(\bar{g} \cdot \hat{B} + \bar{h} \cdot \hat{B}^\perp) + 2(M_1 \bar{h}_1 + M_2 \bar{h}_2)(\hat{M} \cdot \hat{B}^\perp) + 2(M_1 \bar{g}_1 + M_2 \bar{g}_2)(\hat{M} \cdot \hat{B})$$
$$= 2z_1^2 c_1 w_1 + 2z_2^2 c_2 w_2 \ .$$

Hence,

$$J_0 = -4z_1^2 z_2^2 + 2c_1 w_1 z_1^2 + 2c_2 w_2 z_2^2 - c_1 c_2 w_1 w_2$$
$$= -4\Big(z_1^2 - \frac{c_1}{2}w_1\Big)\Big(z_2^2 - \frac{c_2}{2}w_2\Big).$$

and

$$\frac{1}{4}\gamma\big((\gamma + M_3)(M_1^2 + M_2^2) - 2\bar{h}_3(M_1 b_2 - M_2 b_1) + 2\bar{g}_3(M_2 b_2 + M_1 b_1)\big)$$
$$= \gamma(\gamma + M_3)z_1 z_2 + \frac{1}{2}h_3(z_2 c_1 + z_1 c_2) + \frac{1}{2i}g_3(z_2 c_1 - z_1 c_2))$$

Therefore,

$$-\frac{1}{4}J = \Big(z_1^2 - \frac{c_1}{2}w_1\Big)\Big(z_2^2 - \frac{c_2}{2}w_2\Big) - \gamma\Big((\gamma + M_3)z_1 z_2 + \frac{1}{2}h_3(z_2 c_1 + z_1 c_2) + \frac{1}{2i}g_3(z_2 c_1 - z_1 c_2)\Big)$$

is another integral of motion for $\mathcal{H}$. This integral of motion is a slight generalization of the one reported in ([**Bb**]) and coincides with the one found by S. Kowalewski when $h = 0$ and $\gamma = 0$. We remind the reader that $\lambda_3 = \dfrac{1}{2}$ which explains the factor of $\dfrac{1}{2}$ in front of the constants c_1 and c_2.

It turns out that $\left(z_1^2 - \dfrac{c_1}{2} \left(w_1 - \dfrac{c_1}{2} \right) \right) \left(z_2^2 - \dfrac{c_2}{2} \left(w_2 - \dfrac{c_2}{2} \right) \right) - \gamma \Big((\gamma - M_3) z_1 z_2 + \dfrac{h_3}{2} (z_1 c_2 + z_2 c_1) - \dfrac{1}{2} a\bar{a} M_3 \Big)$ is the integral of motion for the Hamiltonian

$$\mathcal{H} = \frac{1}{2}(M_1^2 + M_2^2) + M_3^2 - \gamma M_3 + h_1 b_2 + h_2 b_1$$

on $SO_4(\mathbb{C})$ which suggests that the Kowalewski's integral on $so_4(\mathbb{C})$ is also related to the spectral invariants of some Lie group G. However, such a group is not yet known.

Our findings can be summarized as follows. Let $\mathfrak{g} = so_5(\mathbb{C})$. Then $\mathfrak{g}$ is a ten dimensional Lie algebra of rank 2. Hence generic co-adjoint orbits in $\mathfrak{g}^*$ are 8-dimensional symplectic submanifolds of $\mathfrak{g}^*$. It follows that the maximal number of functionally independent integrals of motion in involution that each left-invariant Hamiltonian could possibly have is equal to four.

The Hamiltonian $2\mathcal{H} = M_1^2 + M_2^2 + 2M_3^2 + 2\alpha M_3 + \hat{B} \cdot g + \hat{B}^\perp \cdot h$ admits $I_0 = M_3 + M_4$ as an integral of motion. Evidently, $\mathcal{H} = I_1$ and I_0 are in involution. It is only on the orbit $I_0 = \gamma = -\alpha$ that there are extra integrals of motion. The reader can easily verify that $I_3 = (g - \hat{B}) \cdot (g - \hat{B}) + (h - \hat{B}^\perp) \cdot (h - \hat{B}^\perp) = (g_1 - b_1)^2 + (g_2 - b_2)^2 + g_3^2 + (h_1 + b_2)^2 + (h_2 - b_1)^2 + h_3^2$ Poisson commutes with I_1 and I_0 on this orbit. Furthermore, one can also show that each of the integrals of motion I_8 and J defined above are in involution with I_1, I_2, I_3 on the orbit $\gamma = -\alpha$. Among these five functions only four are functionally independent, hence $\mathcal{H}$ admits a maximal number of integrals of motion on the orbit $\gamma = -\alpha$ and consequently $\mathcal{H}$ is completely integrable on this orbit.

Kowalewski's integral of motion occurs as a particular case of the above situation when $0 = \gamma = \alpha$ and $h = \hat{B}^\perp = 0$. The remainder of the paper will be devoted to the integration procedure of the Kirchhoff-Kowalewski elastic case.

2. Algebraic preliminaries.

We now shift attention to the algebraic variety V defined by the integrals of motion

$$I_1 = \mathcal{H} = \frac{1}{4\lambda}(M_1^2 + M_2^2) + \frac{1}{2\lambda}M_3^2 + b_1 p_1 + b_2 p_2$$
$$I_2 = p_1^2 + p_2^2 + p_3^2 + \epsilon(M_1^2 + M_2^2 + M_3^2) \, ,$$
$$I_3 = p_1 M_1 + p_2 M_2 + p_3 M_3$$
$$I_4 = \left(z_1^2 - a(w_1 - \epsilon a)\right)\left(z_2^2 - \bar{a}(w_2 - \epsilon \bar{a})\right) \, .$$

We shall simplify the equations by taking $a = 1$ which implies that $b_1 = 1, b_2 = 0$ and $\lambda = 1$. This normalization amounts to rescaling the coordinates and does not change the basic structure of V. We shall also introduce the following variables

$$q_1 = z_1^2 - w_1 + \epsilon \text{ and } q_2 = z_2^2 - w_2 + \epsilon.$$

where $z_i = \frac{1}{2}(M_1 \pm iM_2)$ and $w_i = p_1 \pm ip_2$, $i = 1, 2$. Then

$$I_1 = z_1 z_2 + \frac{1}{2}M_3^2 + \frac{1}{2}(w_1 + w_1)$$
$$I_2 = w_1 w_2 + p_3^2 + \epsilon(4z_1 z_2 + M_3^2)$$
$$I_3 = z_1 w_2 + z_2 w_1 + p_3 M_3$$
$$I_4^2 = q_1 q_2.$$

Now replace w_1 and w_2 by their values $w_1 = z_1^2 - q_1 + \epsilon$ and $w_2 = z_2^2 - q_2 + \epsilon$.

It follows that

$$2(I_1 - \epsilon) = (z_1 + z_2)^2 - (q_1 + q_2) + M_3^2$$
$$I_2 - 2\epsilon(I_1 - \epsilon) - \epsilon^2 = z_1^2 z_2^2 + p_3^2 + 2\epsilon z_1 z_2 - (z_1^2 q_2 + z_2^2 q_1)$$
(77)
$$I_3 = (z_1 z_2 + \epsilon)(z_1 + z_2) - (z_1 q_2 + z_2 q_1) + p_3 M_3$$
$$I_4 = q_1 q_2$$

We shall use $V(\tilde{I}_1, \tilde{I}_2, I_3, I_4)$ to denote the analytic variety (77) where $\tilde{I}_1 = 2(I_1 - \epsilon)$ and $\tilde{I}_2 = I_2 - 2\epsilon(I_1 - \epsilon) - \epsilon^2$ are modified constants of motion.

It is evident that V is invariant under the involution $\sigma(z_1, z_2, w_1, w_2, p_3, M_3) = (z_2, z_1, w_2, w_1, -p_3, -M_3)$. The fixed points of σ given by $z_1 = z_2$, $w_1 = w_2$, $p_3 = M_3 = 0$ also happen to be singular on V. The reader may easily check by using equation (25) that the fixed points of σ correspond to the equilibrium points of our Hamiltonian flow.

PROPOSITION 9.1. $\mathcal{C} \times \mathcal{C}$ is a 4-fold cover of the variety $V(I_1, I_2, I_3, I_4)$.

PROOF. Let

$$\zeta_1 = M_3 z_1 - p_3 \text{ and } \zeta_2 = M_3 z_2 - p_3.$$

Then,

$$\zeta_1^2 = M_3^2 z_1^2 - 2M_3 p_3 z_1 + p_3^2 =$$
$$\big(\tilde{I}_1 + (q_1 + q_2) - (z_1 + z_2)^2\big)z_1^2 - 2\big(I_3 - (z_1 z_2 + \epsilon)(z_1 + z_2) + z_1 q_2 + z_2 q_1\big)z_1 +$$
$$\big(\tilde{I}_2 - z_1^2 z_2^2 - 2\epsilon z_1 z_2 + z_1^2 q_2 + z_2^2 q_1\big) = -z_1^4 + (\tilde{I}_1 + 2\epsilon)z_1^2 - 2I_3 z_1 + \tilde{I}_2 + q_1(z_1 - z_2)^2 =$$
$$- z_1^4 + 2\mathcal{H}z_1^2 - 2I_3 z_1 + \tilde{I}_2 + q_1(z_1 - z_2)^2 = P(z_1) + q_1(z_1 - z_2)^2$$

because

$$(q_1 + q_2)z_1^2 - 2(z_1 q_2 + z_2 q_1)z_1 + z_1^2 q_2 + z_2^2 q_2 = q_1(z_1 - z_2)^2 ,$$

and

$$-(z_1 + z_2)^2 z_1^2 + 2(z_1 z_2 + \epsilon)(z_1 + z_2)z_1 - z_1^2 z_2^2 - 2\epsilon z_1 z_2 = -z_1^4 + 2\epsilon z_1^2 .$$

A similar calculation shows that $\zeta_2^2 = P(z_2) + q_2(z_1 - z_2)^2$.

If we now define the new variables

(79) $$u_1^2 = \zeta_1^2 - q_1(z_1 - z_2)^2 \quad \text{and} \quad u_2^2 = \zeta_2^2 - q_2(z_1 - z_2)^2 ,$$

then

$$u_1^2 = P(z_1) \quad \text{and} \quad u_2^2 = P(z_2).$$

i.e., $M = (z_1, \pm u_1)$ and $N = (z_2, \pm u_2)$ are points on the elliptic curve $u^2 = P(z)$.

The preceding calculations show that each point $(z_1, z_2, w_1, w_2, p_3, M_3)$ of V can be mapped into the product $\mathcal{C} \times \mathcal{C}$ where $\mathcal{C}$ is the elliptic curve $u^2 = P(z)$ via the formulas:

(80) $$u_1^2 = (M_3 z_1 - p_3)^2 - q_1(z_1 - z_2)^2, \ u_2^2 = (M_3 z_2 - p_3)^2 - q_2(z_1 - z_2)^2$$

To show that the mapping (80) is surjective, we need to establish additional algebraic relations.

To each fourth degree polynomial

$$P(z) = A + 4Bz + 6Cz^2 + 4Dz^3 + Ez^4$$

we shall associate a symmetric form $R(z_1, z_2)$ defined by

$$R(z_1, z_2) = A + 2B(z_1 + z_2) + 3C(z_1^2 + z_2^2) + 2D(z_1 + z_2)z_1 z_2 + E z_1^2 z_2^2 .$$

Evidently $R(z, z) = P(z)$.

It is known ([**Ju3**], Theorem 4) that the function

$$F(z_1, z_2) = P(z_1)P(z_2) - R^2(z_1, z_2)$$

is of the form

$$F(z_1, z_2) = (z_1 - z_2)^2 \hat{R}(z_1, z_2)$$

where $\hat{R}$ is a symmetric bi-quadratic form given by

$$\hat{R}(z_1, z_2) = \hat{A} + 2\hat{B}(z_1 + z_2) + 3\hat{C}(z_1^2 + z_2^2) + 2\hat{D}z_1z_2(z_1 + z_2) + \hat{E}z_1^2z_2^2 - \hat{F}(z_1^2 - z_2^2)$$

with

(80)
$$\hat{A} = -4B^2, \quad \hat{B} = 2(AD - 3BC), \quad \hat{C} = \frac{2}{3}(AE + 2BD - 9C^2),$$
$$\hat{D} = 2(BE - 3CD), \quad \hat{E} = -4D^2, \quad \text{and} \quad \hat{F} = 9C^2 - AE - 4BD.$$

In particular $\hat{R}(z, z)$ defines a polynomial $\hat{P}$ given by

(81)
$$\hat{P}(z) = 4\big(-B^2 + 2(AD - 3BC)z + (AE + 2BD - 9C^2)z^2 + 2(BE - 3CD)z^3 - D^2z^4\big).$$

Any bi-quadratic form $R_\theta(z_1, z_2) = R(z_1, z_2) - \theta(z_1 - z_2)^2$ parametrized by a complex number θ satisfies

$$\begin{aligned}
R_\theta^2(z_1, z_2) &= \big(R(z_1, z_2) - \theta(z_1 - z_2)^2\big)^2 \\
&= R^2(z_1, z_2) - 2R(z_1, z_2)(z_1 - z_2)^2\theta + \theta^2(z_1 - z_2)^4 \\
&= P(z_1)P(z_2) - (z_1 - z_2)^2\hat{R}(z_1, z_2) - 2\theta R(z_1, z_2)(z_1 - z_2)^2 + \theta^2(z_1 - z_2)^4 \\
&= P(z_1)P(z_2) - (z_1 - z_2)^2\hat{R}_\theta(z_1, z_2)
\end{aligned}$$

where

(82)
$$\hat{R}_\theta(z_1, z_2) = -\theta^2(z_1 - z_2)^2 + 2R(z_1, z_2)\theta + \hat{R}(z_1, z_2).$$

Let us note the significance of the above relations for the algebraic variety V. The bi-quadratic form $R(z_1, z_2)$ associated to the Kowalewski polynomial is given by

$$R(z_1, z_2) = \tilde{I}_2 - I_3(z_1 + z_2) + \mathcal{H}(z_1^2 + z_2^2) - z_1^2z_2^2.$$

Recall that $\zeta_1 = M_3z_1 - p_3$ and that $\zeta_2 = M_3z_2 - p_3$. Hence

$$(\zeta_1 - \zeta_2)^2 = M_3^2(z_1 - z_2)^2 = \big(\tilde{I}_1 - (z_1 + z_2)^2 + q_1 + q_2\big)(z_1 - z_2)^2.$$

Therefore,

$$
\begin{aligned}
\zeta_1 \zeta_2 &= \frac{1}{2}\left(\zeta_1^2 + \zeta_2^2 - \left(\tilde{I}_1 - (z_1 + z_2)^2 + q_1 + q_2\right)(z_1 - z_2)^2\right) \\
&= \frac{1}{2}\left(u_1^2 + u_2^2 + (q_1 + q_2)(z_1 - z_2)^2 - \left(\tilde{I}_1 - (z_1 + z_2)^2 + q_1 + q_2\right)(z_1 - z_2)^2\right) \\
&= \frac{1}{2}\left(P(z_1) + P(z_2) - \left(\tilde{I}_1 - (z_1 + z_2)^2\right)(z_1 - z_2)^2\right) \\
&= \tilde{I}_2 - I_3(z_1 + z_2) + \epsilon(z_1 - z_2)^2 + 2\mathcal{H}z_1 z_2 - z_1^2 z_2^2 \\
&= \tilde{I}_2 - I_3(z_1 + z_2) + \mathcal{H}(z_1^2 + z_2^2) - z_1^2 z_2^2 - (\mathcal{H} - \epsilon)(z_1 - z_2)^2 \\
&= R(z_1, z_2) - (\mathcal{H} - \epsilon)(z_1 - z_2)^2 \ .
\end{aligned}
$$

The bi-quadratic form $R_0(z_1, z_2) = \zeta_1 \zeta_2$ is of the form R_θ with $\theta = \mathcal{H} - \epsilon$. But then, according to the above there is a unique bi-quadratic form $\hat{R}_\theta$ which, for reasons of notational conformity with [Ju3], we shall denote by $-R_1$ that satisfies

$$
R_0^2(z_1, z_2) = P(z_1)P(z_2) - (z_1 - z_2)^2 R_1(z_1, z_2) = u_1^2 u_2^2 - (z_1 - z_2)^2 R_1(z_1, z_2) \ .
$$

Since

$$
\begin{aligned}
R_0^2 &= \zeta_1^2 \zeta_2^2 = \left(u_1^2 + q_1(z_1 - z_2)^2\right)\left(u_2^2 + q_2(z_1 - z_2)^2\right) \\
&= u_1^2 u_2^2 + \left(u_1^2 q_2 + u_2^2 q_1 + q_1 q_2(z_1 - z_2)^2\right)(z_1 - z_2)^2 \ ,
\end{aligned}
$$

it follows that

(83) $$ R_1(z_1, z_2) + u_1^2 q_2 + u_2^2 q_1 + I_4^2(z_1 - z_2)^2 = 0 \ . $$

The coefficients of $R_1(z_1, z_2)$ can be easily computed explicitly by using formulas (80) (as was done in [Ju3] on page 620). Since the particular expression for R_1 does not figure in the remainder of the paper this detail is omitted.

Relation (83) is fundamental for showing that the smooth points of V are parameterized by points in $\mathcal{C} \times \mathcal{C}$. The argument is as follows:

let (u_1, z_1) and (u_2, z_2) denote any points in $\mathcal{C}$. The substitution of $q_2 = \frac{1}{q_1} I_4^2$ into (83) yields

$$
u_1^2 I_4^2 + u_2^2 q_1^2 + \left(R_1 + I_4^2(z_1 - z_2)^2\right)q_1 = 0 \ .
$$

Hence,

$$
q_1 = -\frac{\left(R_1 + I_4^2(z_1 - z_2)^2\right) + \sqrt{\left(R_1 + I_4^2(z_1 - z_2)^2\right)^2 - 4u_1^2 u_2^2 I_4^2}}{2q_1^2} \ .
$$

Once q_1 and q_2 are determined, then the remaining variables p_3 and M_3 are determined from the relations

$$M_3 z_1 - p_3 = \zeta_1 = \sqrt{u_1^2 - q_1(z_1 - z_2)^2} \,,$$

and

$$M_3 z_2 - p_3 = \zeta_2 = \sqrt{u_2^2 - q_2(z_1 - z_2)^2} \,.$$

3. Euler-Weil Addition formulas.

We now recall another algebraic fact which is needed for proper understanding of the integration procedure used by S. Kowalewski, namely the addition formulas of L. Euler and A. Weil. Consider again the most general symmetric bi-quadratic form $R_\theta(z_1, z_2)$ such that $R_\theta(z, z) = P(z)$. As already stated there is a unique form $\Phi_\theta(z_1, z_2)$ defined by the relation

$$R_\theta^2(z_1, z_2) - (z_1 - z_2)^2 \Phi_\theta(z_1, z_2) = P(z_1)P(z_2) \,.$$

A theorem of L. Euler [**Eu**], whose proof is also given in [**Ju3**], states that $\Phi_\theta(z_1, z_2) = 0$ is a solution of either

$$(84) \qquad \frac{dz_1}{\sqrt{P(z_1)}} + \frac{dz_2}{\sqrt{P(z_2)}} = 0 \,, \quad \text{or} \quad \frac{dz_1}{\sqrt{P(z_1)}} - \frac{dz_2}{\sqrt{P(z_2)}} = 0$$

for each value of θ. Conversely, the solutions of either differential equation correspond to $\Phi_\theta(z_1, z_2) = 0$ for some value θ.

Let us now incorporate A. Weil's algebraic interpretations ([**Wl**]) of the findings of Euler. Since $\Phi_\theta(z_1, z_2)$ is symmetric with respect to its variables, the set of points $\Phi_\theta(z_1, z_2) = 0$ is represented by

$$a_\theta(z_1)z_2^2 + 2b_\theta(z_1)z_2 + c_\theta(z_1) = a_\theta(z_2)z_1^2 + 2b_\theta(z_2)z_1 + c_\theta(z_2) = 0$$

where the coefficients $a_\theta(z)$, $b_\theta(z)$, and $c_\theta(z)$ are quadratic polynomials in both variables θ and z. It was shown in [**Ju3**] that the discriminant $G_\theta(z) = b_\theta^2(z) - a_\theta(z)c_\theta(z)$ is given by $G_\theta(z) = p(\theta)P(z)$ where

$$(85) \qquad p(\theta) = 2\theta(\theta - 3C)^2 + 2\theta(4BD - AE) + 4B^2E + 4AD^2 - 24BCD \,.$$

Then expression (84) is naturally linked with the cubic elliptic curve Γ given by

$$(86) \qquad \eta^2 = 4\xi^3 - g_2\zeta - g_3$$

where g_2 and g_3 are the covariant invariants of $u^2 = P(z)$ given explicitly by

$$g_2 = AE - ABD + 3C^2 \text{ and } g_3 = ACE + 2BCD - AD^2 - B^2E - C^3.$$

We simply take $\theta = 2(\xi + C)$ and define $\eta^2 = \dfrac{p(\theta)}{4}$.

A. Weil points out in [**Wl**] that the theorem of Euler is intimately connected with a group structure on $\mathcal{C} \cup \Gamma$ defined as follows: first note that Γ acts on $\mathcal{C}$. Indeed each point $M = (z, u)$ of $\mathcal{C}$ and each point $\Delta = (\xi, \eta)$ of Γ the corresponding discriminant $G_\theta(z)$ with $\theta = 2(\xi + C)$ is given by $G_\theta(z) = p(\theta)P(z) = 4\eta^2 u^2$.

The relation

$$a_\theta(z)w^2 + 2b_\theta(z)w + c_\theta(z) = 0$$

defines two complex numbers w given by

$$w = -\frac{b_\theta(z) \pm \sqrt{G_\theta(z)}}{a_\theta(z)} = -\frac{b_\theta(z) \pm 2\eta u}{a_\theta(z)}.$$

Let $w_1 = -\frac{b_\theta(z) + 2\eta u}{a_\theta(z)}$, and then let $N = (w_1, u_1)$ denote the point on $\mathcal{C}$ such that

$$z = -\frac{b_\theta(w_1) - 2\eta u_1}{a_\theta(w_1)}, \text{ or } u_1 = -\frac{1}{2\eta}\big(a_\theta(w_1)z + b_\theta(w_1)\big).$$

Following A. Weil we write

$$\Delta + M = N.$$

If we now write $-M = (z, -u)$ for a point $M = (z, u)$ of $\mathcal{C}$ it then follows from above that $\Delta - M = (w_2, u_2)$ where

$$w_2 = -\frac{b_\theta(z) - 2\eta u}{a_\theta(z)} \text{ and } u_2 = -\frac{1}{2\eta}\big(a_\theta(w_2)z + b_\theta(w_2)\big).$$

The zero set $\Phi_\theta(z, w) = 0$ can be expressed as $z(w)$ in some open neighborhood of the point z provided that $G_\theta(z) \neq 0$. It follows that

$$\frac{dz}{dw} = -\frac{\partial \Phi_\theta}{\partial w} \Big/ \frac{\partial \Phi_\theta}{\partial z} = -\frac{2\big(a_\theta(z)w + 2b_\theta(z)\big)}{2\big(a_\theta(w)z + 2b_\theta(w)\big)}.$$

In particular at (z_1, w_1) we have

$$\frac{dz}{dw} = -\frac{a_\theta(z)w_1 + b_\theta(z)}{a_\theta(w_1)z + b_\theta(w_1)} = \frac{u}{u_2}$$

and at (z, w_2)

$$\frac{dz}{dw} = -\frac{a_\theta(z)w_2 + b_\theta(z)}{a_\theta(w_2)z + b_\theta(w_2)} = -\frac{u}{u_2}.$$

In the first case $z(w)$ is a solution of $\dfrac{dz}{\sqrt{P(z)}} - \dfrac{dw}{\sqrt{P(w)}} = 0$, while in the second case $z(w)$ is a solution of $\dfrac{dz_1}{\sqrt{P(z)}} + \dfrac{dw}{\sqrt{P(w)}} = 0$.

Suppose that $(z_1, u_1) = M$ and $(z_2, u_2) = N$. It then follows that the sum $M + N$ and the difference $M - N$ are in Γ for any two points M and N of $\mathcal{C}$. The points $\Delta = (\xi, \eta)$ on Γ that correspond to $M + N$ and $M - N$ are determined by the appropriate values of θ such that $\Phi_\theta(z_1, z_2) = 0$. According to the expression (82)

$$\theta = \frac{R(z_1, z_2) \pm \sqrt{R^2(z_1, z_2) + (z_1 - z_2)^2 \hat{R}(z_1, z_2)}}{(z_1 - z_2)^2}.$$

Since $R^2(z_1, z_2) + (z_1 - z_2)^2 \hat{R}(z_1, z_2) = P(z_1)P(z_2)$, it follows that

$$\theta = \frac{R(z_1, z_2) \pm u_1 u_2}{(z_1 - z_2)^2}.$$

Each choice of θ determines complex numbers $\xi = 2(\theta + C)$ which in turn determines the appropriate point on Γ. The value of θ that corresponds to $\Delta = M + N$ must tend to a finite limit when N approaches M, while the value of θ that corresponds to $\Delta = M - N$ tends to infinity, since Δ tends to the group identity on Γ.

It is easy to see that the correct choice of θ is given $\theta = \dfrac{R(z_1, z_2) - u_1 u_2}{(z_1 - z_2)^2}$.

Indeed,

$$\frac{R(z_1, z_2) - u_1 u_2}{(z_1 - z_2)^2} = \frac{\big(R(z_1, z_2) - u_1 u_2\big)\big(R(z_1, z_2) + u_1 u_2\big)}{\big(R(z_1, z_2) + u_1 u_2\big)(z_1 - z_2)^2}$$

$$= \frac{R^2(z_1, z_2) - u_1^2 u_2^2}{(z_1 - z_2)^2\big(R(z_1, z_2) + u_1 u_2\big)} = -\frac{(z_1 - z_2)^2 \hat{R}(z_1, z_2)}{(z_1 - z_2)^2\big(R(z_1, z_2) + u_1 u_2\big)}$$

$$= -\frac{\hat{R}(z_1, z_2)}{R(z_1, z_2) + u_1 u_2}.$$

As z_2 tends to z_1, the preceding expression tends to $-\dfrac{\hat{P}(z_1)}{P(z_1) + u_1^2} =$ $-\dfrac{\hat{P}(z_1)}{2P(z_1)}$, where $\hat{P}$ denotes the polynomial defined by (81).

Thus $\theta_1 = \dfrac{R(z_1, z_2) - u_1 u_2}{(z_1 - z_2)^2}$ determines $\xi_1 = 2(\theta_1 + C)$, while $\theta_2 = \dfrac{R(z_1 z_2) + u_1 u_2}{(z_1 - z_2)^2}$ determines $\xi_2 = 2(\theta_2 + C)$. Then,

$$M + N = (\xi_1, \eta_1) = \Delta_1, \text{ and } N - M = (\xi_2, \eta_2) = \Delta_2.$$

The values of η_1 and η_2 are uniquely determined through the relations $\Delta_1 - M = N$ and $\Delta_2 + M = N$.

The transformations which appear in Kowalewski's paper correspond to the infinitesimal version of Weil's addition formulas. For if $M = (z_1, u_1)$ and $N = (z_2, u_2)$ denote arbitrary points of $\mathcal{C}$ and $\Delta_1 = (\xi_1, \eta_1)$ and $\Delta_2 = (\xi_2, \eta_2)$ denote the points of Γ such that $-M + N = \Delta_1$ and $M + N = \Delta_2$ then,

$$\frac{d\xi_1}{\eta_1} = -\frac{dz_1}{u_1} + \frac{dz_2}{u_2} , \qquad \text{and} \qquad \frac{d\xi_2}{\eta_2} = \frac{dz_1}{u_1} + \frac{dz_2}{u_2} .$$

With this background material at our disposal we come to the integration procedure of Kowalewski.

4. Integration.

It follows from equations (66) that

$$2\frac{dz_1}{dt} = i(M_3 z_1 - p_3)$$

$$2\frac{dz_2}{dt} = -i(M_3 z_2 - p_3).$$

Hence,

$$-4\left(\frac{dz_1}{dt}\right)^2 = (M_3 z_1 - p_3)^2 = \zeta_1^2 = u_1^2 + q_1(z_1 - z_2)^2$$

$$-4\left(\frac{dz_2}{dt}\right)^2 = \zeta_2^2 = u_2^2 + q_2(z_1 - z_2)^2.$$

Rather than integrating these equations on $\mathcal{C} \times \mathcal{C}$ we shall integrate them on $\Gamma \times \Gamma$ through the transformation

$$M - N = \Delta_1 \quad \text{and} \quad M + N = \Delta_2 .$$

whose infinitesimal analogue is given by

$$\frac{d\xi_1}{\eta_1} = \frac{dz_1}{u_1} - \frac{dz_2}{du_2} , \quad \text{and} \quad \frac{d\xi_2}{d\eta} = \frac{dz_1}{u_1} + \frac{dz_2}{du_2} .$$

Along each extremal curve $z_1(t), z_2(t)$

$$\left(\frac{d\xi_1}{dt}\right)^2 \frac{1}{\eta_1^2(t)} = \frac{1}{u_1^2(t)}\left(\frac{dz_1}{dt}\right)^2 + \frac{1}{u_2^2(t)}\left(\frac{dz_2}{dt}\right)^2 - \frac{2}{u_1(t)u_2(t)}\frac{dz_1}{dt}\frac{dz_2}{dt}$$

and

$$\left(\frac{d\xi_2}{dt}\right)^2 \frac{1}{\eta_2^2(t)} = \frac{1}{u_1^2(t)}\left(\frac{dz_1}{dt}\right)^2 + \frac{1}{u_2^2(t)}\left(\frac{dz_2}{dt}\right)^2 + \frac{2}{u_1(t)u_2(t)}\frac{dz_1}{dt}\frac{dz_2}{dt} .$$

But,

$$\frac{dz_1}{dt} = i\frac{1}{2}\big(M_3(t)z_1(t) - p_3(t)\big) = i\frac{1}{2}\zeta_1(t) , \quad \text{and}$$

$$\frac{dz_2}{dt} = -i\frac{1}{2}\big(M_3(t)z_2(t) - p_3(t)\big) = -i\frac{1}{2}\zeta_2(t) .$$

Hence,

$$\left(\frac{dz_1}{dt}\right)\left(\frac{dz_2}{dt}\right) = \frac{1}{4}\zeta_1(t)\zeta_2(t) = \frac{1}{4}R_0\big(z_1(t), z_2(t)\big) \ .$$

After the substitutions,

$$\frac{-4}{\eta_1^2}\left(\frac{d\xi_1}{dt}\right)^2 = 2 + \big(z_1(t) - z_2(t)\big)^2\left(\frac{q_1(t)}{u_1^2(t)} + \frac{q_2(t)}{u_2^2(t)}\right) - \frac{2R_0\big(z_1(t), z_2(t)\big)}{u_1(t)u_2(t)}$$

and

$$\frac{-4}{\eta_2^2}\left(\frac{d\xi_2}{dt}\right)^2 = 2 + \big(z_1(t) - z_2(t)\big)^2\left(\frac{q_1(t)}{u_1^2(t)} + \frac{q_2(t)}{u_2^2(t)}\right) + \frac{2R_0\big(z_1(t), z_2(t)\big)}{u_1(t)u_2(t)} \ .$$

Therefore,

$$-\frac{4}{\eta_1^2}\left(\frac{d\xi_1}{dt}\right)^2 = \frac{2u_1^2u_2^2 + (z_1 - z_2)^2(u_1^2q_2 + u_2^2q_1) - 2R_0u_1u_2}{u_1^2u_2^2} \ .$$

Since $u_1^2q_2 + u_2^2q_1 + R_1 + I_4^2(z_1 - z_2^2) = 0$ and since $R_0^2 = u_1^2u_2^2 - (z_1 - z_2)^2R_1$, the right hand side of the above equation simplifies to

$$-\frac{4}{\eta_1^2}\left(\frac{d\xi_1}{dt}\right)^2 = \left(\frac{u_1u_2 - R_0}{u_1u_2}\right)^2 - \frac{(z_1 - z_2)^4 I_4^2}{u_1^2u_2^2} \ .$$

Now recall that $\theta = \dfrac{R - u_1u_2}{(z_1 - z_2)^2}$ and that $\theta = 2(\xi + C)$. Also recall that $R_0 = R_\theta$ with $\theta = \mathcal{H} - \epsilon$. Hence, $R_0 = R - (\mathcal{H} - \epsilon)(z_1 - z_2)^2$ and consequently,

$$\frac{R_0 - u_1u_2}{(z_1 - z_2)^2} = \frac{R - u_1u_2}{(z_1 - z_2)^2} - (\mathcal{H} - \epsilon) = \theta - (\mathcal{H} - \epsilon) = 2(\xi + C) - \mathcal{H} + \epsilon$$

$$= 2\xi + \frac{2\mathcal{H}}{3} - \mathcal{H} + \epsilon$$

$$= 2\left(\xi - \frac{\mathcal{H}}{6} + \frac{\epsilon}{2}\right) \ .$$

After the substitution,

$$-\frac{4}{\eta_1^2}\left(\frac{d\xi_1}{dt}\right)^2 = 4\frac{(z_1 - z_2)^4}{u_1^2u_2^2}\left(\left(\xi_1 - \frac{\mathcal{H}}{6} + \frac{\epsilon}{2}\right)^2 - \frac{I_4^2}{4}\right) \ .$$

In a completely analogous manner

$$-\frac{4}{\eta_2^2}\left(\frac{d\xi_2}{dt}\right)^2 = \frac{(u_1u_2 + R_0)^2}{u_1^2u^2} - \frac{(z_1 - z_2)^4}{u_1^2u_2^2}I_4^2 = \frac{4(z_1 - z_2)^4}{u_1^2u_2^2}\left(\left(\xi_2 - \frac{\mathcal{H}}{6} + \frac{\epsilon}{2}\right)^2 - \frac{I_4^2}{4}\right) \ .$$

Let $k_1 = \dfrac{\mathcal{H}}{6} - \dfrac{\epsilon}{6} + \dfrac{I_4}{2}$, and let $k_2 = \dfrac{\mathcal{H}}{6} - \dfrac{\epsilon}{6} - \dfrac{I_4}{2}$.

Then,

$$\left(\frac{d\xi_1}{dt}\right)^2 = \eta_1^2 \frac{(z_1 - z_2)^4}{u_1^2 u_2^2}(\xi_1 - k_1)(\xi_1 - k_2) , \quad \text{and}$$

$$\left(\frac{d\xi_2}{dt}\right)^2 = \eta_2^2 \frac{(z_1 - z_2)^4}{u_1^2 u_2^2}(\xi_2 - k_1)(\xi_2 - k_2) .$$

Since $\xi_2 - \xi_1 = \frac{1}{2}(\theta_2 - \theta_1) = \frac{u_1 u_2}{(z_1 - z_2)^2}$ it follows that

$$\left(\frac{d\xi_1}{dt}\right)^2 = \frac{-\eta_1^2}{(\xi_1 - \xi_2)^2}(\xi_1 - k_1)(\xi_1 - k_2) , \quad \text{and}$$

$$\left(\frac{d\xi_2}{dt}\right)^2 = \frac{-\eta_2^2}{(\xi_1 - \xi_2)^2}(\xi_2 - k_1)(\xi_2 - k_2) .$$

But $\eta_1^2 = 4\xi_1^3 + g_2\xi_1 - g_3$ and $\eta_2^2 = 4\xi_2^3 + g_2\xi_2 - g_3$. So

$$\left(\frac{d\xi_1}{dt}\right)^2 = \frac{U(\xi_1)}{(\xi_1 - \xi_2)^2} \quad \text{and} \quad \left(\frac{d\xi_2}{dt}\right)^2 = \frac{U(\xi_2)}{(\xi_1 - \xi_2)^2}$$

where $U(\xi) = -(4\xi^3 - g_2\xi - g_3)(\xi - k_1)(\xi - k_2)$.

It follows that $\left(\frac{d\xi_1}{d\xi_2}\right)^2 = \left(\frac{U(\xi_1)}{U(\xi_2)}\right)^2$, and therefore $\frac{d\xi_1}{\sqrt{U(\xi_1)}} \pm$
$\frac{d\xi_2}{\sqrt{U(\xi_2)}} = 0$.

The correct sign is determined through the following argument:
Equation $\frac{d\xi_1}{\sqrt{U(\xi_1)}} - \frac{d\xi_2}{\sqrt{U(\xi_2)}} = 0$ admits a solution $\xi_1 = \xi_2$. This
solution impels that $\Delta_1 = (\xi_1, \eta_1) = \Delta_2 = (\xi_2, \eta_2)$, which in turn implies that $M + N = \Delta_1 = \Delta_2 = M - N$. Hence $N = 0$. But 0 belongs to Γ and does not belong to $\mathcal{C}$. Hence

(87)
$$\frac{d\xi_1}{\sqrt{U(\xi_1)}} + \frac{d\xi_2}{\sqrt{U(\xi_2)}} = 0$$

is the correct equation. This equation coincides with the equation of
Kowalewski reported in her celebrated paper of 1889 ([**Kw**]).

The solutions of the last equation are given by $F(\xi_1) + F(\xi_2) = $
constant where F is the hyperelliptic integral

$$F(z) = \int_{z_0}^{z} \frac{d\zeta}{\sqrt{U(\zeta)}} .$$

From a theoretical point of view the above formula provides a complete solution to our Hamiltonian system, although the task of unraveling this answer back to the original variables and to the elastic curves remains a laborious exercise. We will not undertake this task now in detail; instead, we make some simple comparisons between the solution in the Lagrange's case and the solutions in the Kowalewski's case.

5. Curvature and torsion of elastic curves.

We return to the space forms and the curvature and the torsion of the corresponding elastic curves. Recall the Serret-Frenet frames $T(t)$, $N(t)$, $B(t)$ along the curve $\gamma(t)$ are given by the following formulas $\frac{d\gamma(t)}{dt} = T(t)$, $\frac{dT}{dt} = k(t)N(t)$, $\frac{dN}{dt} - k(t)T(t) + \tau(t)B(t)$, and $\frac{dB}{dt} = -\tau(t)N(t)$.

We shall identify γ and the Serret-Frenet frame along γ with a curve $g(t)$ in the isometry group G, as explained in the early part of the paper, in which case $g(t)$ is a solution of the following differential system $\dfrac{dg(t)}{dt} = g(t) \begin{pmatrix} 0 & -\epsilon & 0 & 0 \\ 1 & 0 & -k(t) & 0 \\ 0 & k(t) & 0 & -\tau(t) \\ 0 & 0 & \tau(t) & 0 \end{pmatrix}$.

For the elastic curves in the case of Lagrange, the comparisons between the Serret-Frenet frame and the elastic frame are easy, because both frames are adapted to the curve in such a way that the first leg of the frame coincides with the tangent of the curve. Hence, the two frames rotate in the plane perpendicular to the tangent vector. If $\beta(t)$ denotes the angle by which the two frames differ then

$$k(t) = u_3(t) \cos \beta(t) - u_2(t) \sin \beta(t) \ ,$$

$$\frac{d\beta}{dt} + u_1(t) = \tau(t) \ , \quad \text{and}$$

$$u_2(t) \cos \beta(t) + u_3(t) \sin \beta(t) = 0$$

for any framed curve $\gamma(t) = g(t)e_1$ where

$$\frac{dg(t)}{dt} = g(t) \begin{pmatrix} 0 & -\epsilon & 0 & 0 \\ 1 & 0 & -u_3(t) & u_2(t) \\ 0 & u_3(t) & 0 & -u_1(t) \\ 0 & -u_2(t) & u_1(t) & 0 \end{pmatrix}.$$

For the derivation of this fact, see [**Ju2**], p. 461. In particular, the elastic curves that correspond to the Hamiltonian $\mathcal{H} = \frac{1}{2\lambda_1}M_1^2 + \frac{1}{2\lambda}(M_2^2 + M_3^2) + p_1$ are generated by the extremal controls $u_1(t) = \frac{M_1}{\lambda_1}$, $u_2(t) = \frac{1}{\lambda}M_2(t)$ and $u_3(t) = \frac{1}{\lambda}M_3(t)$. Since $M_1 = $ constant, $u_1(t)$ is constant. Then,

$$\kappa^2 = u_3^2 + u_2^2 = \frac{1}{\lambda^2}(M_2^2 + M_3^2) \ .$$

After rescaling the Hamiltonian so that $\lambda = 1$ we get $\kappa^2 = u_1^2 + u_2^2 = M_2^2 + M_3^2$. Furthermore,

$$\frac{d\beta}{dt} = M_1\left(1 - \frac{1}{\lambda_1}\right) - (M_2 p_2 + M_3 p_3)/(M_2^2 + M_3^2) \qquad (\text{see } [])$$

and therefore

$$\tau = M_1\left(1 - \frac{1}{\lambda_1}\right) - (M_2 p_2 + M_3 p_3)/(M_2^2 + M_3^2) + \frac{1}{\lambda_1}M_1$$

or

$$\kappa^2 \tau = M_1 k^2 - (M_1 p_2 + M_2 p_2 + M_3 p_3) + M_1 p_1 \ .$$

In particular, $\kappa^2(t)\tau(t) = \text{constant}$ when $M_1 = 0$. The elastic curves for which $M_1 = 0$ coincide with the solutions of the Euler-Griffiths problem (Proposition 7.1) The curves that correspond to $I_3 = 0$ satisfy $\tau(t) = 0$ and hence are confined to a surface.

Now turn to the curvature and torsion of the elastic curves associated with the Hamiltonian $\mathcal{H} = \frac{1}{4}(M_1^2 + M_2^2) + \frac{1}{2}M_3^2 + p_1$.

The extremal controls are given by

$$u_1(t) = \frac{1}{2}M_1(t) \qquad u_2(t) = \frac{1}{2}M_2(t) \quad \text{and} \quad u_3(t) = M_3(t) \ .$$

Then the associated curvature $\kappa(t)$ and the torsion $\tau(t)$ are given by the usual formulas

$$\kappa(t) = u_3(t)\cos\beta(t) - u_2(t)\sin\beta(t) \ ,$$

$$\text{subject to} \quad u_3(t)\sin\beta(t) + u_2(t)\cos\beta(t) = 0$$

$$\text{where} \quad \frac{d\beta(t)}{dt} + u_1(t) = \tau(t) \ .$$

Hence,

$$\kappa^2(t) = u_2^2(t) + u_3^2(t) = \frac{1}{4}M_2^2(t) + M_3^2(t) \ .$$

After differentiating $\tan\beta(t) = -\frac{1}{2}M_2(t)/M_3(t)$ we get

$$\kappa^2\frac{d\beta}{dt} = \frac{1}{2}\left(M_3\left(p_3 - \frac{1}{2}M_1 M_3\right) + M_2 p_2\right) = -\frac{1}{2}I_3 + \frac{M_1}{2}\left(\frac{M_3^2}{2} + p_1\right) \ ,$$

and therefore

$$\kappa^2\tau = \kappa^2 u_1 + \kappa^2\frac{d\beta}{dt} = \kappa^2\frac{M_1}{2} - \frac{1}{2}I_3 + \frac{M_1}{2}\left(\mathcal{H} - \frac{1}{4}(M_1^2 + M_2^2)\right)$$

$$= -\frac{1}{2}I_3 + \frac{1}{2}M_1\left(\mathcal{H} - \frac{1}{4}M_1^2 + M_3^2\right) \ .$$

Apart from the geodesics which correspond to the fixed points of the Hamiltonian flow $(z_1 = z_2, w_1 = w_2, p_3 = M_3 = 0)$, there are no other solutions for which $\tau(t) = 0$ i.e., there are no other planar curves.

Consider now the particular solution for which

$$q_1 = z_1^2 - w_1 + \epsilon = 0 \text{ and } q_2 = z_2^2 - w_2 + \epsilon = 0.$$

This is possible since $\dfrac{dq_1}{dt} = iM_3(t)q_1(t)$ and $\dfrac{dq_2}{dt} = -iM_3(t)q_2(t)$. Then $-4\left(\dfrac{dz_1}{dt}\right)^2 = P(z_1)$, and that $-4\left(\dfrac{dz_2}{dt}\right)^2 = P(z_2)$ with $P(z) = \tilde{I}_2 - 2I_3 z + 2\mathcal{H}z^2 - z^4$.

The solutions of $\dfrac{dz}{dt} = \sqrt{P(z)}$ with $P(z)$ a fourth degree polynomial describe the arc-length of lemniscate, which according to C.L. Siegel [**Sg**] was a principal motivation for Euler's interest in this differential equation.

The assumption $q_1 = q_2 = 0$ implies that the algebraic variety V reduces to a curve since the variables z_1 and z_2 are related through $R_1(z_1, z_2) = 0$ (equation (83)).

We shall conclude this section with a differential equation for the curvature of an elastic curve. Parametrize κ by the points on curve $u^2 = P(z)$. Then,

$$\frac{d}{dt}\kappa^2(t) = 2\kappa(t)\frac{d\kappa}{dt} = 2\kappa(z)\frac{d\kappa}{dz}\frac{dz}{dt}$$

and so

$$4\kappa^2\left(\frac{dz}{dt}\right)^2\left(\frac{d\kappa}{dz}\right)^2 = 4\kappa^2 P(z)\left(\frac{d\kappa}{dz}\right)^2.$$

But

$$\frac{d}{dt}\kappa^2 = \frac{1}{2}M_2(t)\dot{M}_2(t) + M_3\dot{M}_3 = \frac{1}{2}M_2\left(p_3 - \frac{1}{2}M_1M_3\right) - M_3p_2\ .$$

So

$$\begin{aligned}
4\kappa^2\left(\frac{dk}{dt}\right)^2 = 4\kappa^2 u^2\left(\frac{d\kappa}{dt}\right)^2 &= \left(\frac{1}{2}M_2\left(p_3 - \frac{1}{2}M_1M_3\right) - M_3p_2\right)^2 \\
&= \frac{1}{4}M_2^2\left(p_3 - \frac{1}{2}M_1M_3\right)^2 + M_3^2p_2^2 - M_2M_3p_2\left(p_3 - \frac{1}{2}M_1M_3\right) \\
&= \frac{1}{4}M_2^2\left(p_3^2 + \frac{1}{4}M_1^2M_3^2 - p_3M_3M_1\right) + M_3^2p_2^2 - M_2p_2M_3p_3 + \frac{1}{2}M_1M_2M_3^2p_2\ .
\end{aligned}$$

We now recall the integrals of motion

$$\tilde{I}_1 = (z_1 + z_2)^2 - (q_1 + q_2) + M_3^2 = (z_1 + z_2)^2 + M_3^2$$

$$\tilde{I}_2 = z_1^2 + z_2^2 + p_3^2 + 2\epsilon z_1 z_1 - (z_1^2 q_2 + z_2^2 q_1) = z_1^2 z_2^2 + p_3^2 + 2\epsilon z_1 z_2$$

$$I_3 = (z_1^2 + \epsilon)(z_1 + z_2) - (z_1 q_2 + z_2 q_1) + p_3 M_3 = (z_1 z_2 + \epsilon)(z_1 + z_2) + p_3 M_3 \ .$$

In addition $q_1 = z_1^2 - w_1 + \epsilon = 0$, $q_2 = z_2^2 - w_2 + \epsilon = 0$.

Hence,

$$p_3 M_3 = I_3 - (z_1 z_2 + \epsilon)(z_1 + z_2), \ M_3^2 = \tilde{I}_1 - (z_1 + z_2)^2$$

$$p_3^2 = \tilde{I}_2 - z_1^2 z_2^2 - 2\epsilon z_1 z_2, \ p_2 = \frac{1}{2i}(z_2^2 - z_1^2).$$

Of course, $M_1 = z_1 + z_2$ and $M_2 = -i(z_1 - z_2)$.

Substitution of these values in the above differential equation for $\kappa(t)$ yields

$$4\kappa^2 P(z_1) \left(\frac{dk}{dz_1}\right)^2 = Q(z_1, z_2)$$

with $Q(z_1, z_2)$ a homogeneous polynomial of degree four in the variables z_1, z_2. Therefore, $\kappa^2 \left(\dfrac{d\kappa}{dz}\right)^2$ is a rational function of the independent variable z induced by the relation $R_1(z_1, z_2) = 0$.

Bibliography

[Ar1]   V.I. Arnold, *Mathematical Methods of Classical Mechanics*, Graduate Texts in Mathematics, vol. 60, Springer-Verlag, New York, 1978.

[Ar2]   V.I. Arnol'd and A.B. Givental', *Symplectic Geometry and its Applications*, Encyclopaedia of Mathematical Sciences (editors V.I. Arnol'd and S.P. Novikov) **4** (1990), Springer-Verlag, Berlin Heidelberg, 4-136.

[Bb]   A.I Bobenko, A.G. Reyman and M.A. Semenov-Tian Shansky, *The Kowalewski top 99 years later; a Lax pair, generalizations and explicit solutions*, Comm. Math. Phys **122** (1989), 321-354.

[Bo]   O. Bogoyavlenski, *Integrable Euler equations on SO_4 and their physical applications*, Comm. Math. Phys **93** (1984), 417-436.

[Ca]   E. Cartan, *The Theory of Spinors*, Hermann, Paris, 1966.

[dC]   M..P.do Carmo, *Riemannian Geometry*, Mathematics:Theory and Applications, Birkhäuser, Boston, 1992.

[Eu]   L. Euler, *Evolutio generalior formularum comparationi curvarum inserventium*, Opera Omnia Ser I^a **20** (**E347/1765**), 318-354.

[Go]   V.V. Golubev, *Lectures on the Integration of the Equations of Motion of a heavy body around a fixed point*, (in Russian), Gostekhizdat, Moscow, 1977.

[Gr]   P. Griffiths, *Exterior Differential Systems and the Calculus of Variations*, Birkhäuser, Boston, 1983.

[Ha]   L. Haine, *Geodesic Flow on $SO(4)$ and Abelian Surfaces*, Math. Ann. **263** (1983), 435-472.

[Hg]   S. Helgason, *Differential Geometry and Symmetric Spaces*, Academic Press, New York, 1962.

[HvM]   E. Horozov and P. Van Moerbeke, *The full geometry of Kowalewski's top and $(1,2)$ Abelian surfaces*, Comm. Pure and App. Math. **XLII** (1989), 357-407.

[IvS]   C. Ivanescu and A. Savu, *The Kowalewski top as a reduction of a Hamiltonian system on $Sp(4,\mathbb{R})$*, Proc. Amer. Math. Soc. **131** (2003), 607-618.

[Ju1]   V. Jurdjevic, *Non-Euclidean Elasticae*, Amer. J. Math. **117** (1995), 93-125.

[Ju2]   V. Jurdjevic, *Geometric Control Theory*, Studies in Advanced Math, vol. 52, Cambridge Univ. Press, New York, 1997.

[Ju3]   V. Jurdjevic, *Integrable Hamiltonian Systems on Lie groups: Kowalewski type*, Annals Math. **150** (1999), 605-644.

[JuM]   V. Jurdjevic and F. Perez-Monroy, *Variational problems on Lie groups and their homogeneous spaces: elastic curves, tops and constrained geodesic problems*, Contemporary Trends in Nonlinear Control Theory and its Applications; (editors B. Bonnard et al) (2002), World Scientific, Singapore, 3-51.

[Ki]   A.A. Kirillov, *Unitary representations of nilpotent Lie groups*, Usp. Mat.Nauk **17, No 4** (1962), 57-110.

[Kw]   S. Kowalewski, *Sur le problème de la rotation d'un corps solide autor d'un point fixé*, Acta Math. **12** (1889), 177-232.

[Ko]   I. V. Komarov, *Kovalewski top for the hydrogen atom*, Theor. Math. Phys **47(1)** (1981), 67-72.

[KoK]   I. V. Komarov and V. B. Kuznetsov, *Kowalewski top on the Lie algebras $o(4)$, $e(3)$ and $o(3,1)$*, J. Phys. A **23(6)** (1990), 841-846.

[Kz]   V.V. Kozlov, *Nonexistence of single-valued integrals and branching of solutions in rigid body mechanics*, Prikl. Mat. Mekh **42** (1978), no. 3, 400-406.

[LaS]   J. Langer and D. Singer, *Knotted elastic curves in IR^3*, J. London Math. Soc. **2, 30**
           (1984), 512-534.

[Lvl]   R. Liouville, *Sur le movement d'un corps solide pesant suspendue par l'un de ces points*,
           Acta Math. 20 (1896), 239-284.

[LuS]   W. Liu and H.J. Sussmann, *Shortest paths for sub-Riemannian metrics on rank 2
           distributions*, Amer. Math. Soc. Memoirs (564) (1995).

[Lv]   A.E. Love, *A Treatise on the Mathematical Theory of Elasticity*, 4th edition, Dover,
           New York, 1927.

[Ly]   A.M. Lyapunov, *On a certain property of the differential equations of a heavy rigid
           body with a fixed point*, Soobshch. Kharkov. Mat. Obshch. Ser. 2, 4 (1894), 123-140.

[Mo]   J. Moser, *Regularization of Kepler's problem and the Averaging Method on a Manifold*,
           Comm. Pure Appl. Math. **XXIII** (1970), 609-636.

[Ra]   T. Ratiu, *The motion of the free n-dimensional Rigid body*, Indiana Univ. Math J. **29**
           (1980), no. 4, 602-629.

[RSTS]   A. G. Reyman and M. A. Semenov-Tian Shansky, *Group-theoretic methods in the the-
           ory of finite-dimensional integrable systems*, Encyclopaedia of Mathematical Sciences
           (edited by V. I. Arnold and S.P. Novikov), Part 2, Chapter 2 (1994), Springer-Verlag,
           Berlin Heidelberg.

[Sg]   C.L. Siegel, *Topics in Complex Function Theory*, vol. 1; *Elliptic Functions and Uni-
           formization Theory,*, Tracts in Pure and Appl. Math. **25** (1969), Wiley-Interscience,
           John Wiley and Sons, New York.

[St]   S. Sternberg, *Lectures on Differential Geometry*, Prentice Hall Inc., Englewood Cliffs,
           N.J., 1964.

[Su]   H.J. Sussmann, *Geometry and Optimal Control*, Mathematical Control Theory (eds J.
           Baillieul and J.C. Willems) (1998), Springer- Verlag, New York, 240-298.

[Wl]   A. Weil, *Arithmetic and Geometry*, Euler and the Jacobians of elliptic curves, vol. I;
           Progr. Math. **35** (1983), Birkhäuser, Boston, 353-359.

[Zg1]   S.L. Ziglin, *Branching of solutions and Non-Existence of first integrals in Hamiltonian
           Mechanics I*, Funkts. Anal. Prilozhen **16** (1982), no. 3, 30-41.

[Zg2]   S.L. Ziglin, *Branching of solutions and Non-Existence of first integrals in Hamiltonian
           Mechanics II*, Funkts. Anal. Prilozhen **17** (1981), no. 1, 8-23.

Editorial Information

To be published in the *Memoirs*, a paper must be correct, new, nontrivial, and significant. Further, it must be well written and of interest to a substantial number of mathematicians. Piecemeal results, such as an inconclusive step toward an unproved major theorem or a minor variation on a known result, are in general not acceptable for publication. Papers appearing in *Memoirs* are generally at least 80 and not more than 200 published pages in length. Papers less than 80 or more than 200 published pages require the approval of the Managing Editor of the Transactions/Memoirs Editorial Board.

As of July 31, 2005, the backlog for this journal was approximately 14 volumes. This estimate is the result of dividing the number of manuscripts for this journal in the Providence office that have not yet gone to the printer on the above date by the average number of monographs per volume over the previous twelve months, reduced by the number of volumes published in four months (the time necessary for preparing a volume for the printer). (There are 6 volumes per year, each containing at least 4 numbers.)

A Consent to Publish and Copyright Agreement is required before a paper will be published in the *Memoirs*. After a paper is accepted for publication, the Providence office will send a Consent to Publish and Copyright Agreement to all authors of the paper. By submitting a paper to the *Memoirs*, authors certify that the results have not been submitted to nor are they under consideration for publication by another journal, conference proceedings, or similar publication.

Information for Authors

Memoirs are printed from camera copy fully prepared by the author. This means that the finished book will look exactly like the copy submitted.

The paper must contain a *descriptive title* and an *abstract* that summarizes the article in language suitable for workers in the general field (algebra, analysis, etc.). The *descriptive title* should be short, but informative; useless or vague phrases such as "some remarks about" or "concerning" should be avoided. The *abstract* should be at least one complete sentence, and at most 300 words. Included with the footnotes to the paper should be the 2000 *Mathematics Subject Classification* representing the primary and secondary subjects of the article. The classifications are accessible from `www.ams.org/msc/`. The list of classifications is also available in print starting with the 1999 annual index of *Mathematical Reviews*. The Mathematics Subject Classification footnote may be followed by a list of *key words and phrases* describing the subject matter of the article and taken from it. Journal abbreviations used in bibliographies are listed in the latest *Mathematical Reviews* annual index. The series abbreviations are also accessible from `www.ams.org/publications/`. To help in preparing and verifying references, the AMS offers MR Lookup, a Reference Tool for Linking, at `www.ams.org/mrlookup/`. When the manuscript is submitted, authors should supply the editor with electronic addresses if available. These will be printed after the postal address at the end of the article.

Electronically prepared manuscripts. The AMS encourages electronically prepared manuscripts, with a strong preference for $\mathcal{AMS}$-LaTeX. To this end, the Society has prepared $\mathcal{AMS}$-LaTeX author packages for each AMS publication. Author packages include instructions for preparing electronic manuscripts, the *AMS Author Handbook*, samples, and a style file that generates the particular design specifications of that publication series. Though $\mathcal{AMS}$-LaTeX is the highly preferred format of TeX, author packages are also available in $\mathcal{AMS}$-TeX.

Authors may retrieve an author package from e-MATH starting from `www.ams.org/tex/` or via FTP to `ftp.ams.org` (login as `anonymous`, enter username as password, and type `cd pub/author-info`). The *AMS Author Handbook* and the *Instruction Manual* are available in PDF format following the author packages link from `www.ams.org/tex/`. The author package can be obtained free of charge by sending email to pub@ams.org (Internet) or from the Publication Division, American Mathematical Society, 201 Charles St., Providence, RI 02904, USA. When requesting an author package, please specify $\mathcal{AMS}$-LaTeX or $\mathcal{AMS}$-TeX, Macintosh or IBM (3.5) format, and the publication in which your paper will appear. Please be sure to include your complete mailing address.

Sending electronic files. After acceptance, the source file(s) should be sent to the Providence office (this includes any TeX source file, any graphics files, and the DVI or PostScript file).

Before sending the source file, be sure you have proofread your paper carefully. The files you send must be the EXACT files used to generate the proof copy that was accepted for publication. For all publications, authors are required to send a printed copy of their paper, which exactly matches the copy approved for publication, along with any graphics that will appear in the paper.

TeX files may be submitted by email, FTP, or on diskette. The DVI file(s) and PostScript files should be submitted only by FTP or on diskette unless they are encoded properly to submit through email. (DVI files are binary and PostScript files tend to be very large.)

Electronically prepared manuscripts can be sent via email to pub-submit@ams.org (Internet). The subject line of the message should include the publication code to identify it as a Memoir. TeX source files, DVI files, and PostScript files can be transferred over the Internet by FTP to the Internet node e-math.ams.org (130.44.1.100).

Electronic graphics. Comprehensive instructions on preparing graphics are available at www.ams.org/jourhtml/graphics.html. A few of the major requirements are given here.

Submit files for graphics as EPS (Encapsulated PostScript) files. This includes graphics originated via a graphics application as well as scanned photographs or other computer-generated images. If this is not possible, TIFF files are acceptable as long as they can be opened in Adobe Photoshop or Illustrator. No matter what method was used to produce the graphic, it is necessary to provide a paper copy to the AMS.

Authors using graphics packages for the creation of electronic art should also avoid the use of any lines thinner than 0.5 points in width. Many graphics packages allow the user to specify a "hairline" for a very thin line. Hairlines often look acceptable when proofed on a typical laser printer. However, when produced on a high-resolution laser imagesetter, hairlines become nearly invisible and will be lost entirely in the final printing process.

Screens should be set to values between 15% and 85%. Screens which fall outside of this range are too light or too dark to print correctly. Variations of screens within a graphic should be no less than 10%.

Inquiries. Any inquiries concerning a paper that has been accepted for publication should be sent directly to the Electronic Prepress Department, American Mathematical Society, 201 Charles St., Providence, RI 02904, USA.

Titles in This Series

TITLES IN THIS SERIES

For a complete list of titles in this series, visit the
AMS Bookstore at **www.ams.org/bookstore/**.